P9-DVO-534

BIOCODE

BIOCODE

THE NEW AGE OF GENOMICS

DAWN FIELD & NEIL DAVIES

OXFORD
UNIVERSITY PRESS

Great Clarendon Street, Oxford, OX2 6DP,
United Kingdom

Oxford University Press is a department of the University of Oxford.
It furthers the University's objective of excellence in research, scholarship,
and education by publishing worldwide. Oxford is a registered trade mark of
Oxford University Press in the UK and in certain other countries

First Edition published in 2015
Impression: 1

Published in the United States of America by Oxford University Press
198 Madison Avenue, New York, NY 10016, United States of America

British Library Cataloguing in Publication Data
Data available

Library of Congress Control Number: 2014950232

ISBN 978–0–19–968775–6

Printed in Great Britain by
Clays Ltd, St Ives plc

This book is dedicated to
Ryan, Suse, John, Felipe, Mesude, and Katja (DF),
and to Maheata (ND)

PREFACE

DNA is a biological code elegantly composed of only four letters: A, C, G, and T. From this simplicity comes all the complexity of life. The key message of this book is that despite all the tremendous achievements the era of genomics is only starting. We are still seeing just the earliest, fuzziest glimmers of deep insight compared to the richness of life on Earth and the questions we can use it to answer. We stand on the cusp of sequencing the Earth from genome to ecosystem, from our own guts to our oceans.

In the course of eight chapters we attempt to span the breadth of the study of genomics from the discovery of the DNA double helix to the impending promise of planetary-scale genomics. Breakthroughs came in thick and fast during this project and we hope that the breadth of topics helps convey how fast the field is moving. A complete set of endnotes and references provides links to further reading or use in the classroom.

We need to thank many people. Top of the list is Latha Menon, editor extraordinaire. She shepherded this book through all steps from first enquiry to published form, often graciously sharing her wisdom over coffee. Likewise, we are indebted to her assistant Emma Ma for her help and advice and to Oxford University Press for making the project possible.

Critical readers are like gold dust and Suse (and John!) Field was a fountain of excellent feedback throughout the project. Andrew Singer was the best possible sounding board for early ideas of the 'genomic stories' we might best select. Jack Gilbert and Rob Knight merit special kudos for reading and commenting on complete drafts—but above all for doing so from the perspective of being colleagues driving so much of the science that is shaping the emerging field of 'biodiversity genomics'. Hillary Gilbert and Martha Beeman provided outstanding proofing and comments. We are appreciative to many others for engaging with us by reading and discussing, including Lita Procter, Adrienne Minock, C. J. Cramer, Larry Minock, Robert Possee, Gina Crivello, Peter Sterk, Antonio Fernandez-Guerra, the Monterey Writers' Group, and Aidan Hansell.

All omissions or mistakes are solely ours. The scientific passion, excitement, and achievements of this field are the contributions of the scientists who have published pioneering papers in the field of genomics since its inception in 1995. Recognition must also go to those who have funded this work, and Neil would personally like to thank the Gordon and Betty Moore Foundation for their support of research in Moorea. We appreciate the help of all those who helped us improve our portrayals of their science, including Nikos Kyrpides, Ollie Ryder, Martin Blaser, Paul Hebert, Scott Edmunds, Eske Willerslev, Camilo Mora, George Roderick, David Liittschwager, Chris Meyer, George Roderick, Jonathan Coddington, Leslie Lyons, Eric Alm, Lawrence David, Joakim Larsson, Robert May, Rob Dunn, James Ostheimer, Nick Loman, Larry Smarr, Morten Allentoft, Linda Amaral-Zettler, Nick Loman, Noah Fierer, Karen E. Nelson, Hans-Peter Klenk, Jonathan Eisen, Owen White, Stuart Kim, Gary Wolf, Mike Snyder, Heather Dewey-Hagbord, Tim Smyth, and Nick Goldman. We also thank the hundreds of scientists of the Moorea Biocode Project, the Genomic Standards Consortium, the Genomic Observatories Network, Ocean Sampling Day, and the Moorea Avatar Project for working with us over the past years to make these initiatives possible. Combined, you are all de facto leaders of the unannounced Planetary Genome Project.

DAWN FIELD AND NEIL DAVIES
United Kingdom and French Polynesia, July 2014

CONTENTS

LIST OF ILLUSTRATIONS

1

DNA

Immortal coil

In 1869, Friedrich Miescher published a quietly received paper entitled 'On the chemical composition of pus cells'. It contained his account of isolating a substance that he called 'nuclein'. Miescher, a Swiss doctor, made the discovery while searching for proteins in white blood cells from patient bandages. He had found the fourth major class of cellular molecules, after proteins, fats, and sugars. It would take almost a century, until 1953 to be precise, for scientists to describe the chemical structure of nuclein, by then relabelled deoxyribonucleic acid, or DNA. With this information, science would finally crack the 'secret of life'.

'WE WISH to suggest a structure for the salt of deoxyribose nucleic acid (D.N.A.)' hardly promises a riveting read, but so began one of the most momentous scientific papers ever published.[1] James Watson and Francis Crick's consciously unassuming article followed Miescher's tradition for understatement, but the co-authors fully understood the significance of their breakthrough. As Crick wrote to his 12-year-old son in a handwritten letter, they had found the copying mechanism 'by which life comes from life'.

For the 60th anniversary of the double helix paper, Michael Crick auctioned off his father's letter. The publicity material of Christie's auction house read: 'More than one month before the first published announcement, Francis Crick, the co-discoverer of the structure and function of DNA, details one of the most important scientific discoveries of the 20th century—the "Secret of Life"—to his son.' The lot became the most expensive epistle in history—it sold for US$5.3 million.

In seven pages, Crick, a future Nobel Laureate, provides a scientific explanation of his accomplishment: 'Now we believe that the D.N.A. is a code. That is, the order of the bases (the letters) makes one gene different from another gene (just as one page of print is different from another).' Crick describes the discovery as 'beautiful'. One of the pages has a simple sketch of DNA's double helix structure and the letter is signed 'Lots of love, Daddy.' Watson and Crick had started a revolution: within 50 years the human genome was sequenced, heralding what many consider the 'century of biology'.

Nobel Prizes are bestowed on advances that change our view of the universe and our place in it. Sometimes the transformative nature of the discovery is immediately apparent. More often, the full impact blossoms over time. Choosing the double helix as worthy of a Nobel Prize was a good bet; the rewards associated with human possession of this profound knowledge continue to accrue. Today, we are witnessing the advent of personal, or individualized, genomics and the creation of synthetic life. We stand in the best position yet to reap rich rewards.

The double helix is embedded in us all and in our culture. The immortal coil[2] has long escaped the ivory tower and is increasingly accessible via over-the-counter services and through the creative outputs of artists, fashion designers, musicians, and writers. Thoughts on DNA are expressed carefully in the most profound works of philosophy, and, on the lighter side, across a wide range of media from pop songs to cartoons. While we rewrite the landscape of biology through laboratory research on DNA, the molecule weaves itself deeper into our social fabric.

Salvador Dalí pioneered DNA-as-art. He celebrated the 10th anniversary of Watson and Crick's discovery with *Galacidalacidesoxyribonucleicacid*, a painting whose title mixes his wife's name and an earlier version of the chemical name for DNA. In the painting, his wife, Gala, looks up to the heavens and Dalí mused that the double helix is 'the only structure linking man to God'.[3] Soraya de Chadarevian, the historian of science, put it thus in an article for *The Economist*: 'DNA has taken on an almost sacred status as the central blueprint for life, and the double helix has become the instantly identifiable secular equivalent of a modern religious icon.'[4] DNA was there from the beginning and it connects all life—two characteristics historically reserved for deities.

A world in your wardrobe

DNA has a profound ability to encode information: enough to build even the most complex organism. It is also capable of encoding any other type of information. Technologists are mimicking the exceptional storage properties of DNA to store digital data. There is no gene for Dalí's *Galacidalacidesoxyribonucleicacid* or any other work of art, but DNA is being explored as a type of molecular hard-drive.

The first book to be 'printed' in DNA was created in 2012. Fittingly, it was a book on synthetic biology, *Regenesis: How Synthetic Biology Will Reinvent Nature and Ourselves,*[5] co-authored by genomics pioneer George Church of Harvard University. In what was at the least a magnificent marketing gimmick, Church used an algorithm to translate over 53,000 words and 11 images from digital form into the four-letter language of genetics; he then fabricated the corresponding DNA molecule using the technologies of synthetic genomics. The contents of Church's book, heavy in the palm of the hand, in DNA are smaller than a dust speck.

It is not likely that DNA books will ever replace hard copy or digital books. Nevertheless, we can only imagine what entrepreneurs will come up with if, or some would say when, DNA sequencers are as accessible as smart phones and we invent DNA printers.

DNA storage might be a way to keep pace with 'big data'. Just considering the output of scientific research alone, information storage needs are escalating rapidly. Data is streaming in from environmental sensors, particle physics experiments, and image processing, to name but a few. DNA sequencing machines, for example, are churning out ever-growing quantities of information that must be stored in a global public database if all are to benefit from the new genetic knowledge.

We have a global molecular library of DNA mirrored in the United States, Europe, and Japan. In its first decade, it grew from almost 700,000 base pairs in 1982, to over 148 trillion.[6] If we imagine the sequences aligned in one long strand, the resulting DNA molecule would have stretched 1 per cent of the way to the sun in 1982. By 2012, it would be long enough to reach the sun and back—10,000 times over.[7]

Public DNA holdings are currently doubling every nine months, a remarkable rate that seems set to fall to five months soon.[8] Genetic sequences are just one of the types of information on the planet. Worldwide digital storage overtook the amount of information stored on paper and other analogue media for the first time in 2002. By 2007, 94 per cent of our information was digital, and in 2011, we stored more than 300 exabytes of information globally. One exabyte has 18 zeros after it. This means that there were over 300 times more digital bytes on our hard drives than grains of sand in the world. Or to think about it another way: if a star were considered 1 byte of data, then we have a galaxy of digital data for each person on Earth.[9]

We are now into zettabytes (10^{21} bytes). For those responsible for archiving human knowledge, these statistics are at once exciting and terrifying. At the current rate, it will soon be impossible to archive every cultural artefact we produce. DNA might be part of the solution. Church and colleagues achieved a storage density of 700 terabytes of information per gram of DNA—six orders of magnitude more dense than contemporary computer hard-disks. In 2013, a research team led by Nick Goldman and Ewan Birney of the European Bioinformatics Institute (EBI) outside Cambridge, UK masterminded a new approach that upped storage capacity to 2.2 petabytes.[10]

Importantly, their algorithm was able to reduce errors to such an extent that they could sequence, or 'read' it back out again, with close to 100 per cent accuracy. Showing that the information could be read back with minimal loss of fidelity was a crucial proof of concept. They wrote and read a variety of digital icons, including a photograph of the institute where they work, a text file with all of Shakespeare's Sonnets, a 26-second audio snippet of Martin Luther King's 'I Have a Dream' speech, and a PDF of Watson and Crick's paper describing the DNA double helix.

DNA offers a future-proof technology. One challenge for storage media, as anyone who has tried to get data from a floppy disk could testify, is ensuring that the information format is still decipherable to people, even thousands of years from now. DNA is the language of life; knowledge of how to read and write it will never go out of fashion. Kept cold, dry, and dark, an airtight test tube on a nice safe shelf is enough to store properly prepared DNA for hundreds to thousands of years without substantial degradation. After all, it is possible to get genetic information from woolly mammoths frozen in ice 10,000 years ago.

One gram of DNA can hold half a million DVDs' worth of information, or about 2 petabytes. This is 27 years of high-definition video. The current contents of the US Library of Congress could be stored 50 times over in a teaspoon. Thus a 'pinch of DNA', dried as a white powder, would offer an intriguing new way to consume the world's culture; the DNA edition of the Library of Congress would taste salty, not sweet.

Cells still read and write DNA far better than we do and DNA synthesis is still prohibitively expensive: synthesizing the DNA in Goldman and Birney's modest experiment was performed by the company Agilent and cost about US$20,000. A second obstacle is searching across a body of information stored in DNA. There is no equivalent of a Google search that would find a given text in the molecular Library of Congress—not yet anyway. The best use of DNA storage at the moment still seems to be for long-term archiving; for example, to put all human knowledge into 1,000-year 'time capsules' for posterity—perhaps safe in orbit around the Earth. The sum of current human knowledge could be stored in about 500 kilograms of DNA.[11] Perhaps C. S. Lewis was right after all, and whole worlds might one day be kept in the back of a wardrobe.

The molecular narcissist

DNA sequences of any substantial length are nearly infinitely variable. Even though there are only four building blocks paired along the double helix—A, C, T, and G—the potential variation in their order is virtually endless. Furthermore, each human genome is reshuffled every generation as those of our parents mix in novel ways when sperm and egg come together. As a result of this recombination, no two people share a genome sequence, except for identical siblings.

Each of us has our moniker written inside our cells. All our names start with *Homo sapiens* but then they diverge; our individual identity is revealed long before reading the 3 billion letters that make up our genome. This variability in our DNA is a treasure trove for unique identification—the ultimate fingerprint. Researchers, governments, and entrepreneurs alike are mining and exploiting our DNA diversity in myriad ways. Innovations range from the sublime to the ethically worrisome and the downright disturbing. Others are just fun: Decorating your apartment? How about a rug emblazoned with your dog's 'DNA portrait'?

'DNA-as-art' companies are capitalizing on the genotyping technologies at the heart of molecular biology. They want us to stamp our uniqueness on everything from mugs to wedding bands to private jets. The images are nothing like the ones we are used to seeing in portrait galleries. Rather, they are linear, minimalist sets of horizontal rectangles that nobody would associate with a person unless they knew it was a DNA profile. A radical new way to view ourselves is precisely what seems to appeal.

The DNA portraits offered are most often based on the use of two Nobel Prize winning discoveries: restriction enzymes and the polymerase chain reaction (PCR). The resulting technologies enable us to tap into the endless genetic variation found among humans, whether for forensics, medicine, or art. Restriction enzymes are special proteins found in bacteria. They act as genetic wire cutters, chopping DNA at specific sequences.

Treating DNA with these enzymes yields fragments of different length. When run out under an electrical current across a chunk of a jelly-like substance called agarose, the fragments move at different speeds according to their sizes and the bands of DNA separate. These bands are the rungs of the ladders in the images. Since different genomes have different sequences, the enzymes cut at different locations, yielding characteristic banding patterns that reflect the genetic differences between individuals.

PCR adopts the machinery of life to copy DNA. A highly simplified version can be done in a test tube with just a polymerase, the enzyme that copies DNA, and individual bases, A, C, T, G and short stretches of DNA, called primers, that bookend exactly which short stretch of DNA should be copied.[12] From even a single copy of DNA, billions of copies can be produced, called 'amplifying'. It is an exponential process as one piece of DNA is made into two, which then form the templates for another set of copying reactions leading to four copies and so on. About 40 cycles is usually enough to enable easy manipulation of the amplified DNA. With enough copies of a gene, we can easily manipulate it, for example to read its code or put it into another organism, thus pasting and cutting it, or 'recombining' it, with a different organism.

Together, PCR and the ability to cut and paste DNA constitute the so-called 'recombinant DNA' technologies. Developed in the late 1980s, they underpinned the explosive growth of the biotechnology industry. DNA is changing business models everywhere, and it is having a profound influence on our culture. Much as some now regard the brain as the seat of the modern soul, our genomes are becoming central to our concept of 'self'.

'Who's Your Daddy?'

The unique code within us is a boon for a range of personal identification services. These DNA identity tests hold great promise, but also risk playing havoc with our relationships, insurance industries, and legal systems. Increasingly access to public DNA sequencing has already exposed shocking cases of forged identifies. One case involved a father who found out that his child was not his biological offspring.[13] Investigations found that a male receptionist at a fertility clinic the couple had used had substituted his sperm for that of the husband's. Since this discovery, the Reproductive Medical Technologies Clinic in Salt Lake City has encouraged all couples who used their services between 1986 and 1995 to also submit to DNA testing.

Human DNA testing is becoming so widely accessible that there are now mobile labs roaming major US cities offering a range of DNA services. A pioneer of this concept is the 'Who's Your Daddy?' truck.[14] Brainchild of Jared Rosenthal, founder of Health Street, the truck offers a range of kerbside services including drug screening and background checks, as well as DNA-based tests of relatedness or kinship, including paternity testing. This innovative service was launched in New York in 2010 and quickly spread to other major cities. When in 2013 the truck made its debut in Boston 1,000 people lined up.

The nature of reproductive biology often makes it quite difficult to be sure who has fathered a particular offspring. This inherent uncertainty of fatherhood has important evolutionary consequences across the animal kingdom. In species considered to be monogamous, like many birds for example, the father should be obvious. Many studies, however, reveal a startling number of what biologists dispassionately refer to as 'extra-pair copulations'. Such goings-on are not entirely unknown in human society either, but their prevalence has been difficult to measure—until now. Roadside services, and perhaps one day the routine screening of babies, can make the paternity of offspring immediately evident. The broad availability of such data could have far-reaching, and possibly disruptive, social consequences. It will put to the test as never before vows of marital fidelity.

DNA testing goes far beyond paternity testing. Matching fathers to offspring is just one of the many relationships between humans that can be established by comparing DNA profiles. The Health Street website, for example, highlights a number of stories. Some are heart-warming, others upsetting. Among the former are DNA tests to support immigration requests for relatives or to find long-lost children or parents. On the disturbing end of the spectrum are a married couple who discovered they shared the same father.

The universality of DNA and associated genetic machinery—a lingua franca of life—means that most of the DNA technologies developed for humans are immediately applicable to other species. As a consequence, we are now applying them to the animals and plants we love best: those domesticated species we share our lives with—or eat.

Most US states ban wolf–dog hybrids as being too aggressive and unpredictable to serve as pets. Might a litter of puppies result from your pet's illicit tryst with a wolf? This could sound far-fetched and yet it is estimated that there are more than 300,000 wolf–dog hybrids around today. To find out, a swab from a puppy's cheek can be sent to labs such as the Veterinary Genetics Laboratory (VGL)[15] at University of California Davis, a world-leading animal genotyping laboratory. Scientists have studied the DNA of enough dogs and wolves now to know the differences.

What goes for dogs goes for cats. While there isn't much fear of cat–tiger or cat–lion hybrids, there are plenty of interesting things to learn from your tabby's DNA. According to the Cat Ancestry Test website, developed by Leslie Lyons, most of the 50–60 breeds of cats are less than 100 years old and all cats can be traced back to eight geographic regions of origin: Western Europe, Egypt, East Mediterranean, Iran/Iraq, Arabian Sea, India, South Asia, and East Asia. She has used samples from cat shows held all over the world to create a cat DNA database of ~170 DNA markers that can be used to assign all 29 of the major fancy cat breeds and major geographic regions. Now you can resolve whether your cat is truly from the alley, or if it harbours a more exotic bloodline: Persian, Abyssinian, or—if a bit large—Maine Coon.

Other ingenious applications of DNA testing include checking food quality and nabbing litterbugs. When worried about the contents of hamburger meat, submit your sample to a DNA test for the presence of 12 mammalian species, including horse and dog as well as cow. If you are mad about the dog mess in your town, perhaps you should follow the example of communities that are mandating the entry of neighbourhood dogs into a local canine DNA database. They can now identify the owners who fail to scoop.[16]

Plant and animal breeding is now relying more and more on DNA. Analysis of DNA is one reason why milk yield in cows has improved so significantly over recent years. DNA can also tell whether a male stud will produce horned or hornless offspring. Mating any female with a stud carrying two copies of the *pulled* gene guarantees horn-free offspring, a highly desirable trait in cows, sheep, and big animals kept in high numbers in tight quarters.

One of the goals of genomics is to link genetic sequence variation to physical, mental, and behavioural characters. Knowing the 'genetic quality' of economically important animals (and plants) is paramount, especially given the strains on global food security. Leaving livestock to choose their own mates was abandoned at the dawn of the agricultural revolution, as farmers sought to optimize traits like milk yield and numbers of eggs laid through selective breeding programmes. Livestock today are conceived by artificial insemination and DNA analysis is emerging as the power tool for finding the best genes. Using DNA means supplementing, or even abandoning, years of successful 'gut instinct'. We are quickly moving into an era where all our farmed food will be the result of comparisons and calculations of the DNA kind.

The case of the unusual cat

Every new technology needs a 'killer application', though DNA testing is perhaps the first sector to take that business adage so literally. The first widespread use of DNA profiling for personal identification was in the domain of forensics. Today, DNA testing is a staple of popular TV shows like *CSI: Crime Scene Investigation*. The benefit of using

DNA-based testing as judicial evidence is now firmly established. Analysis of the most variable regions of the human genome from tiny amounts of biological material found at crime scenes can provide a unique match to the perpetrator. Based on past successes the use of DNA testing continues to ramp up and in many countries is now blending into the broader concept of genetic surveillance.

Many American states, for example, now maintain a growing database of genetic profiles that has proven a powerful investigative tool for solving crimes. Use of DNA-based identification was thought to be the same as matching tattoos to known gang members to establish criminal affiliations. Just like fingerprinting or photographing it is legitimate. Civil liberties campaigners in the US were concerned about the breach of privacy, however, and the question ended up before the US Supreme Court in 2013. In the case of *Maryland* v. *King*, the Court ruled to uphold the routine collection of DNA samples from criminal suspects.[17] The Court had been asked to decide whether a recently collected DNA profile could be used to convict the arrestee for his six-year-old crime. The justices ruled by a narrow 5–4 majority that the analysis of an arrestee's DNA is consistent with the US Constitution's Fourth Amendment prohibiting unreasonable searches and seizures.

The ruling risked opening a Pandora's box. Dissenting justices in the minority report agreed that DNA testing could help to solve cold cases, but were concerned about its use to flag potential suspects. While it might be reasonable to 'search the DNA' of someone accused of a serious crime, the definition of a serious crime can be a slippery slope. Might police be able to collect DNA from those committing traffic violations in the future?

Another issue is more intellectually intriguing, and goes to the heart of DNA's power. What about the rights of relatives to privacy? The genotyping of someone under arrest could be considered a reasonable search of that person, but it could conceivably lead to the conviction of a sibling who has not been accused of anything. The genetic signatures of close relatives are naturally very similar, so when police check a suspect's DNA against cold cases, a strong but partial match can suggest that one of the suspect's relatives might have been involved. In other words, by gaining access to one person's DNA for one specific crime—a reasonable search according to the Court—law enforcement could effectively rifle through the metaphorical trunks of every car in that person's extended family. Blind fishing trips are often considered an unreasonable breach of privacy, even though they have the potential to solve tough cases.

It is not only your family that might inadvertently land you in jail; your pet might turn informer too. The first use of cat DNA in a murder case was sparked in 2013 by the discovery of eight hairs from a cat named Tinker on a curtain wrapped around a victim at a crime scene on a beach in Hampshire, England.[18] Initial comparisons with existing feline DNA sequences showed no matches with 493 American cats. A database of a 152 British cats from around the UK was then created, also revealing no matches. The jury agreed. The defendant, the owner of Tinker, was found guilty and given a life sentence.

Stranger visions

We leave DNA behind everywhere we go. Just like cats and other mammals, we shed DNA in our hairs. We slather DNA on anything we put our lips to or touch with our fingertips. Heather Dewey-Hagborg fashions DNA portraits[19] but her faces are created from DNA clues we leave in our wakes (see Figure 1). Scavenging lost hairs and discarded cigarette butts from the sidewalks of New York City, she uses DNA analysis to profile people's physical traits. The slightest source of genetic trash yields ample DNA for analysis. She presages what can be learned from the analysis of anonymous DNA; her art is a statement of the future of genetic surveillance.

Dewey-Hagborg analyses her DNAs at the Genspace lab, a non-profit organization dedicated to promoting education in molecular biology for both children and adults.[20] For a reasonable subscription

Figure 1. New York artist's Heather Dewey-Hagborg DNA-inspired human portraits of strangers. Lack of a detectable SRY gene means the face is female; another DNA test suggests eye colour. Maternal mitochondrial Haplogroup gives information on ethnic appearance. With information genes she creates a 3D face. In the future whole genome sequences could be mined to create holographic profiles of individuals from 'genetic trash', like hairs, left behind in public places.

rate, members of the public can enter the Genspace lab and conduct molecular biology experiments, including the analysis of DNA. Genspace is a major player in the growing movement to democratize DNA analysis technologies. If you want to know if there is horsemeat in your burger, or if your fish is really tuna, find out here. This is the future of DIY DNA.[21]

Traditional artists work from subjects, Dewey-Hagborg from molecules. She uses genes for specific traits like eye colour and gender to create her faces. For example, if the sample has the sex-determing region Y (SRY) gene that determines sex and is found on the Y chromosome, she knows she has a male. A skull would be better for guessing how prominent someone's chin is or how large the nose might be. With a 3D printer she builds each genetically-inspired face into the world. She relies on her imagination to a large extent; this is still art.

Some may find these DNA faces creepy and disturbing. Critics believe her 'Stranger Visions' collection is an intrusion of privacy. But that is part of Dewey-Hagborg's intrigue with the subject. It reveals how much your genome says about you—and how much it does not, at least not yet. The speculative portraits are reminders of how much information humans can inadvertently leave behind. Pinning any sequence to its owner would require a reference database containing their personal genetic profile—or enough of their relatives to make the inference. Just combing your hair in public could put your genomic legacy on display.

Dewey-Hagborg is scraping the surface of what can be learned from DNA. Today, she could just as easily use whole genome sequencing and obtain hundreds of traits—and is working now to do so. This raises an interesting question about the future use of DNA in law enforcement.

While the mainstay of DNA-testing in the judicial domain has long been identification, one can now imagine 3D reconstruction perpetrators. Studies are now showing that genetic tests, known as forensic DNA phenotyping, can indeed prove reliable predictors of hair and eye colour even when applied to the long-deceased, and computer programs now exist to make crude 3D facial representations—'DNA mugshots'.[22]

Advertising your genes

The power of DNA lies in our ability to read out truths about our identity, ancestry, and traits. Genomes are often described as autobiographies. This power will grow as we shift from genetic tests that look at small bits of our genomes towards genome-wide interpretations of 'self'.

DNA genotyping is already disruptive, but complete genomic analysis will make it more so. How much depends on the extent that we pool the data from our genomes. The willingness to share is a key factor. It has nothing to do with the science per se but everything to do with human psychology and motivations. Under what circumstances would you share your DNA profile or full genome sequence? With your family or your doctor only? With people you shared a particular trait with—a disease perhaps? What if it could be credibly de-identified so that in anonymous form it could be mined for scientific research?

We put photos of ourselves on websites and elsewhere as part of our place in society. Perhaps it will be so for our genomes too. Our face gives clues to our gender, age, health, mood, aesthetics, and ancestry. Our genomes do the same, albeit sometimes with more or less detail. Can you imagine sharing your genome as easily as you post pictures on Facebook? How might such a genomic society be different? How will this knowledge change us?

Companies like Google and Amazon are famous for their advertising, customized-to-user keywords and online behaviours. What if these companies had access to your genome? What if the marketing industry knew whether or not you are lactose-intolerant or at high risk of going bald? One company betting on such a future is Miinome.[23] It wants to make genomic information 'actionable', offering you advice based on the analysis of your genes, which you could translate into positive actions.

The founders of Miinome, James Ostheimer, a data scientist, and Paul Saarinen, its CEO, expect public buy-in once there are applications beyond medicine. This includes the promise of looking for new associations, say, between certain genes and a taste for spicy foods. Finding such associations, especially if they are low frequency, will only be possible once more people chip in their genomic profiles. These companies hope for a tipping point, after which people will gladly place their genome into the community pot because it becomes something that 'everyone does'.

To build such a genetic marketplace will take a massive increase in the number of people with access to genome sequencing and willing to share the results. The former seems inevitable, the latter less so. It will require a fundamental shift in our thinking about privacy and huge uptake of new technologies. Current trends suggest it could happen. In 50 years we have gone from the discovery of the double helix to the mapping of the human genome sequence; imagine what the next 50 could bring if only half as productive. The stage indeed seems well set for a 'century of biology'.

2

PERSONAL GENOMICS

Science rock star genomes

> We are here to celebrate the completion of the first survey of the entire human genome. Without a doubt, this is the most important, most wondrous map ever produced by humankind.
>
> President Clinton on the unveiling of the Draft Human Genome Sequence, 2001, White House[24]

Two modern-day explorers, Francis Collins and Craig Venter, stood in the East Wing to present their map. They represented the public- and private-led efforts that had raced to sequence the human genome—and finally settled for a draw.[25] Almost 200 years earlier, Lewis and Clark had stood in the same spot to present their first map of the United States to Thomas Jefferson. This time, instead of a map of a new country, we had a map of ourselves.

It had been an international endeavour. As President Bill Clinton unveiled the human genome sequence, UK Prime Minister Tony Blair was connected via satellite. Parallel celebrations were being staged around the world. The Human Genome Project was hailed as one of the greatest, collaborative scientific achievements ever. At a cost of US$3 billion, it was biology's first 'Big Science' project.

Maps have changed the world. The genomic map laid out the chemical instructions for human life: a digital rendition made up of 3 billion letters.[26] It was hoped that the human genome would help us chart a new course towards better health, just as the map of Lewis and Clark had opened up the western two-thirds of the US.

The explosion of human genome sequencing that has occurred since is in large part thanks to the fact that this 'wondrous map' was openly shared for all humanity. Researchers around the world could pore over its every detail at their leisure. It was available for download from public databases supported by the international governments who funded the sequencing.

The digital genome is presented as a string of A's, C's, G's, and T's, shorthand for the names of the four nucleotides of DNA: adenosine, cytosine, guanine, and thymidine. The first reference versions of the human genome were not from one individual; they were mosaics, collages of the DNA of several people. Both the public and private human genome sequencing projects deemed this necessary to increase representative coverage of diversity and dodge some thorny privacy issues.

In 2007, the first genome of a single and identifiable individual was published.[27] It marked a bold new era of 'genomic pride'. The perceived need for *anonymity* was cast aside and replaced with a vision of people sequencing their genomes for personal as well as public benefit. The introduction of this digitized genome into the public domain was radical because everyone knew whose it was—Craig Venter's. It was a scientific milestone and a social awakening.

This genome was so much more than a sequence. It had a face, body, health record, personality, ancestry—and it was already a celebrity. Venter regularly appears in lists of the most influential people in the world. If science has the equivalent of rock stars, Craig Venter might be Mick Jagger. Just as the Rolling Stones seem to have been on tour since the invention of rock music, Venter has been strutting his stuff on the genomics stage for about as long as genomics has existed.

Venter is best known for founding the company Celera and leading the privately funded project to sequence the human genome. He launched the era of genomics by sequencing the first bacterial genome in 1995. Prior to that, he led the sequencing of gene transcripts, or expressed sequence tags (ESTs).

After such achievement, some might prefer to bask in glory and retire to a well-earned yacht. Venter went for the yacht, but he was far from done with science. He turned *Sorcerer II* into a research vessel. In the best traditions of Cook's *Endeavour* or Darwin's *Beagle*, he took *Sorcerer II* on a voyage of discovery, sequencing marine microbes from the world's oceans and helping to launch the field of metagenomics. He has since become a pioneer of synthetic genomics, setting his sights on engineering new life forms to solve the energy crisis. He is also hoping to find signs of life on Mars. He suggested in his book *Life at the Speed of Light*[28] that we could beam any Martian sequences found back to Earth and 'print out' the extraterrestrial organisms using 3D printers.

Some might think that in sequencing himself first, Venter committed an unscientific act of vanity. Others respect and take comfort from the fact that he turned this new technology on himself. This follows a fine scientific tradition, such as Jonas Salk who injected himself with his own polio vaccine. Whatever your view, Venter made a very personal contribution to science: his genome. It paved the way for others to do likewise: to sequence our genomes in the hope of improving our personal health and prosperity.

The study of Venter's DNA included several advances. Human cells contain two copies of our genome, twin sets of 23 chromosomes: one from our mother and one from our father. Previous projects only sequenced one set to make matters simpler. Venter's team disentangled the two copies. He could see the alleles, or gene variants, given to him by both his mother and his father. The Venter genome was also a bargain. The Human Genome Project topped out at US$3 billion and took 13 years to complete. The price tag of Venter's genome was US $100 million as costs associated with maturing sequencing technology plummeted.

The price of the next human genome to be sequenced was a mere US$1.5 million.[29] Again, the subject of the study was another scientific 'rock star': Nobel Laureate James Watson, the co-discoverer of the double helix. In 2008, the company that launched the 'second generation' of DNA sequencing machines—the 454 sequencing platform—

presented Watson with a DVD containing his genome during a ceremony at the Baylor College of Medicine in Houston, Texas.

None of us like to be considered just a statistic and genomics underscores further how, and perhaps even why, one size does not fit all. Beta blockers, prescribed to Watson for his high blood pressure, for example, mysteriously put him to sleep although they work well for others. After his genome revealed he was more sensitive than most to the drug, he was given a lower-than-average dose; dozy side effects solved. Knowing how drugs work with particular genes is one of the most promising applications of genomics, known as pharmacogenomics. It enables the more efficient targeting of drugs to those who will respond best to them.[30]

At the age of 79 Watson was vocal about the merits of sequencing the long-lived. He feels that perhaps it might be more rewarding to sequence old people than young. What about 1,000 smokers who lived to the age of 100, he asks? You might just find a protective factor. The Scripps Translational Science Institute in San Diego, California helped develop this approach. In 2007, it launched the 'Wellderly Study' to investigate 'the genetic architecture of exceptional healthspan'—those who have lived to beyond 80 in good health.[31]

Of the roughly 20,000 genes in the published map of Watson's genome, one is missing. The aspect of sequencing his genome that worried him the most was whether it would reveal a propensity to dementia. Watson's mind was clear as a bell, but his grandmother had succumbed to Alzheimer's disease. Many consider losing one's mental faculties a personal hell, and perhaps a Nobel Laureate might fear it more than most. Watson decided to hide the sequences of his apolipoprotein E gene, or ApoE, even from himself.

The ApoE gene encodes for a 299 amino acid long lipoprotein that transports important molecules like fat-soluble vitamins and cholesterol into the blood. Of the three main genetic variants, carriers of two copies of the e4 variant have up to 20 times the risk of developing Alzheimer's. Moreover, the mean age of onset of Alzheimer's in people with one copy of e4 is 76. In those with two copies, the age drops to 68. While up to 40–65 per cent of Alzheimer's patients have at least one copy of the e4 allele, around one-third are e4 negative and some with two copies of e4 never even develop the disease. As is often the case, the picture of genetic risk is complex.

It is easy to sympathize with Watson. There is little that can be done to prevent Alzheimer's, although research continues. Some focus more on the peace of mind a test could bring—what if Watson had the best possible ApoE sequences? Wouldn't that be a relief? Genotyping can bring 'bad news' but the knowledge can also enable preventative actions. It can also bring the relief of learning that one is 'genetically healthy' and help avoid unnecessary fear. On the other hand, having a test available but without much hope of avoiding a disease, confronts people with a fear they might otherwise not have dwelled upon.

Most of us would prefer to avoid psychological roulette. Yet, these are the kinds of decision we might have to face more often. In many cases, the choices will not be as stark as they currently are for Alzheimer's. It is far from clear that we are adequately prepared as a society for dealing with the complex issues surrounding our genomic health. Just having to make such choices will create more stress. In some cases, the decision anguish might be worse, overall, than the suffering the tests can prevent. With luck, solutions for Alzheimer's and other such diseases will save us from such dilemmas. Indeed, the more strongly linked a disease is to a gene, the greater the chance that we will discover the mechanism and develop effective therapies. Either way, more of us will choose to look at our genomic maps as the price of sequencing a whole human genome drops below US$1,000.[32]

'Six billion base pairs for six billion people'

Many genomes have been sequenced since those of Venter and Watson, but few initiatives have been as openly ambitious about changing society as the Personal Genome Project,[33] or PGP, spearheaded by George Church of Harvard. The PGP tackled issues of genomic privacy head-on by asking willing volunteers to make all of their genomic data, health records, and other trait data freely available from the start. The field of genomics is shaped not only by understanding DNA, but also by how to act on that knowledge ethically.

'Six billion base pairs for six billion people' had a nice ring to it, Church wrote in his 2005 paper 'The Personal Genome Project'.[34] He was referring to the fact that the seeds of the PGP were sown decades before. He hatched the concept with his mentor Wally Gilbert, one of the inventors of DNA sequencing. Church had helped get the 3 billion-dollar human genome project off the ground, but always felt the price tag was too high. He has been working ever since on his vision of bringing down the cost enough to make genome sequencing accessible to everyone.

Since the dawn of sequencing in 1976, Church and Gilbert believed a large and appealing leap would be to go from his new method for sequencing very short segments of DNA (about 30 letters at a time) to a method to get everyone's full genome sequenced (3 billion letters at a time). Church now finds reluctance to exit the pre-genetic age and move into an era of genome-driven medicine simply unforgivable. Failing to use genomic information is like doing surgery without knowing anatomy.

Church conceived of the PGP as the broadest of medical studies. Among many associated ethical issues, two spectres loomed large over such projects: privacy and insurance. Church's university, Harvard, ordered him to do what he would do unto others to himself first. Church's genome was thus at the start of the PGP queue. Reminiscent of the hi-tech villainy of a Bond film, Church took the code name 'PGP-1', but for far from nefarious reasons. Unlike the criminal geniuses of the SPECTRE organization that want to remain in the shadows, Church is promoting extreme transparency. Such openness demonstrates his good intentions, even if not all are comfortable with his science.

In the summer of 2007, Church launched his dream project by getting nine other volunteers to donate their blood and saliva.

Church's PGP-10 cohort includes high-flyers like best-selling author and psychologist Steven Pinker. Like Watson, Pinker also chose not to look at the prophecy of his ApoE gene.[35] The PGP-10 shared their DNA sequences, medical records, and other personal information on the Web. By 2013, more than 1,000 individuals had enrolled in the 'PGP-1K' and work is ongoing towards 100,000 genomes. While volunteers can withhold their names, with such a lot of data about them going into a public database it is quite likely they could be identified. In its 'open consent' policy, the PGP emphasizes that only those 'who are comfortable sharing their data without any promises of privacy, confidentiality or anonymity' should participate.

Church knows that genomics will only serve early adopters until it goes viral. When only a few have complete genomes there is little information about contents. With wide adoption comes a network effect, not unlike what transpired with the telephone, the fax machine, or the World Wide Web. At first connections are limited by the number of people able to participate in the network. The utility of the service is constrained commensurately. When only a few wealthy types had a mobile phone, it was not much use to the rest of the world. The benefits of having your genome sequenced are similarly limited if few others have been sequenced. Understanding what genes are associated with what traits, including diseases, advances more rapidly the larger the sample available for study. Soon, the benefits—and risks—will become evident.

A scant 30,000 genomes

Writing in 2013, Elizabeth Silverman, Wall Street pundit and author of *Genomics: An Investor's Guide*, lamented we only had 30,000 sequenced human genomes.[36] What shocks most is that she calls this huge number *scant*. Silverman argues that the limited number of genomes available for study stymies the growth of the biotech sector. To cope, the industry is moving towards Genomics 2.0, or population genomics, a transition she dubs 'biotechnology's oldest next big thing'.

What motivates companies to invest in genome sequencing? Even as sequencing becomes affordable, the cost of analysing many genomes remains very expensive. Silverman explains that companies have no choice. There is no alternative but to sequence more genomes. While our genomes are in some respects exceedingly similar, they differ in a few key places that determine significant biological outcomes, for example, why some women have such a high chance of getting breast cancer. Our ability to find the differences and to study their effects depends on large sample sizes.

The first genome was a magnificent map but too many important places were missing. A few critics have lambasted the Human Genome Project for failing to deliver immediate health breakthroughs. Silverman argues it was naive to ever think that it could. Some in the biotech industry might have placed expensive bets on finding examples of genes that caused disease. Such bull's-eyes make it easy to develop treatments, drugs, and preventative measures. But we are more complex creatures.

Confounding the search for a simple answer is the fundamental way in which genomes are organized. The assumption of a single gene basis for disease falls short. Genes work together in complex networks, often with long pathways from sequence to disease—from genotype to phenotype. Genes often act in concert to cause or prevent disease, which is the most likely explanation of why not all individuals carrying BRCA breast cancer mutations develop breast or ovarian cancer. As shown by the higher frequency of lung cancer in long-term smokers, lifestyle and environment can play a larger role in disease than genetics. Most often, however, it is a complex interaction of genes and environment.

On the heels of the Human Genome Project, sequencing projects focused on understanding genetic differences among groups of humans. Specifically, these studies are designed to uncover the associations between genetic signatures and key traits. They add specific locations to the map and link them to functions we care about. Which

DNA variants at what locations in the genome are always found in cohorts of people sharing a particular trait?

The first effort to pioneer a large-scale, collaborative project on the back of the Human Genome Project was the International HapMap Project: 'a catalogue of common genetic variants that occur in human beings'.[37] We are mostly the same along vast tracts of our genome—we are all human after all—but there are key differences. Any two people are 99.5 per cent identical at the genetic level, so it's the 0.5 per cent that explains why we're not all the same. The challenge is to understand which of those differences make some of us sick.

Like many studies of genetic variation, the project focuses on special sites in the genome called 'single nucleotide polymorphism' or SNPs, pronounced 'snips'. SNPs occur at places in the genome where one of the letters varies among humans. There are thought to be about one SNP for every 300 or so nucleotides on average. Most of the variants seem to have no major effect, and are relatively neutral for our health. Just talking at the gross level, it turns out that 90 per cent of James Watson's 2 million SNPs are of the type most commonly found in the human population. Yet more than 200,000, about 10 per cent, of his variants were novel; these are the locations in his genomes that help make him uniquely 'Watsonesque'.

Human genetic diversity is vast. While HapMap is a pillar of our current ability to link particular parts of the genome with specific phenotypes, it has still only explained a drop in the ocean of human genetic diversity. There are now a burgeoning number of genome-wide association studies (GWAS) trying to link traits with variation in the genome. To date, about 10 million SNPs have been identified across the human genome,[38] as well as other types of variation, such as 'copy number variants' (CNVs). These are recorded in the public databases but only a fraction are yet associated with biological traits.

We are now filling in the gaps in our understanding. The Venter and Watson genomes represent white Europeans. To complement those sequences, the first diploid genome of an Asian was published in 2010. Unlike those of Venter and Watson, this genome was of an anonymous male Han Chinese individual with no known genetic diseases. Han Chinese individuals account for nearly 30 per cent of the human population but it raises the question 'What does the rest of the world look like?'

The next genomes out of the queue came from individuals from Nigeria and Korea. In 2011, the company Complete Genomics announced it had 69 complete human genomes in its database.[39] Also that year, the Faroe Islands—a self-governing territory within the Kingdom of Denmark—announced it wanted to become the first nation to offer full human genome sequencing to each of its 50,000 citizens, a project known as FarGen.[40] In 2012, the 1,000 Genomes Project Consortium published a paper entitled 'An integrated map of genetic variation from 1,092 human genomes' and in 2013 reanalysed this data set with respect to drawing insights about cancer.[41]

We can all have our genomes sequenced now. In 2013, the most famous personal DNA sequencing company in the world, Silicon Valley-based 23andMe, boasted a customer base of over 400,000. Until then, 23andMe offered profiling for 240 phenotypes associated with known SNPs, and collected hundreds of fields of optional personal information. The company also offered features in their genomic reports that included an original music soundtrack of your DNA.

The 23andMe website was nurturing a growing community of the genomically literate, a social network that could speak DNA, conversing, sharing results, and learning together. A genomic 'Facebook' does not seem at all far-fetched once one has explored this community. Nor is it just a playground for the elite; as of 2013, anyone with Internet access and ready to pay the US$99 sequencing fee could take part.

How far consumer genetic testing will go remains to be seen. Regulation is a minefield through which genomic entrepreneurs must tread carefully. In 2013, 23andMe received a letter from the US Food and Drug Administration (FDA) forcing them to 'immediately discontinue marketing the PGS [Saliva Collection Kit and Personal Genome Service] until such time as it receives FDA marketing authorization for the device'.[42] The problem lies in using DNA testing for predicting disease.[43] The company was also slapped with a class action lawsuit brought through a court in California.[44]

Shortly after, 23andMe rebranded itself as purely a DNA ancestry company.[45] As of early 2014, the 23andMe service was summarized on its homepage as: 'Find out what your DNA says about you and your family. Trace your lineage back 10,000 years and discover your history from over 750 maternal lineages and over 500 paternal lineages.' No doubt this is not the end of the story. It represents one of the first skirmishes in determining how best to regulate whole-genome sequencing platforms and the personal genomics industry.

Genomics is set to revolutionize health care—from the top down as well as individuals up. In 2012, UK Prime Minister David Cameron announced the formation of Genomics England, a company set up by the Department of Health to sequence 100,000 human genomes.[46] The first 8,000 are to be completed in 2014 followed by 30,000 genomes per year thereafter. This project aims to bolster the UK's National Health Service (NHS). One of the advantages of having a medical system that includes just about everyone in the country is the massive database it provides. As the UK's public system bears the brunt of insuring as well as treating patients, it has ample incentive to invest in potential preventative approaches that can reduce costs as well as making people healthier. Genomic medicine offers such hope, especially when large samples are available for data mining. The UK effort brings genome sequencing a step closer as part of routine health care services.

New genomic projects routinely eclipse one another in size and coverage, somewhat like the competition to build the world's tallest building. The need for ever taller buildings might be questioned, but we know that even today's large-scale genomic studies address but the tiniest slice of human genetic diversity. We can expect to see the size and shape of genomic projects continue to grow well into the coming decade. If the brief history of genomics is any guide, today's large projects will appear exceedingly tame tomorrow.

Genomics 101

Preparations are being made to sequence the next generation. Just as this generation of kids is comfortable with a smart phone, the next will be familiar with their genomes. The signs are everywhere. The isolation of DNA is an increasingly common science-fair demonstration for kids. Fortunately, we can do without Miescher's pus-filled bandages. A much more pleasant source of DNA is fruit, especially strawberries or bananas. Easily mashed, even by small fingers if slipped into a closed Ziploc bag, just add a bit of washing-up liquid and the cells break. Then douse with high-percentage alcohol and the double helix crystallizes, or precipitates, just like table salt.

Translucent globs of stringy, snot-like DNA appear in seconds. The grown-up version of this trick can include using high-proof rum. Video instructions for 'DNA cocktails' are on YouTube. If you are adventurous and want to see your own DNA, you can expel enough spit into a glass to get sufficient cells (go for about an inch of liquid). After precipitating the DNA into long silvery threads, one can twirl it up on a toothpick, like using a stick to sweep the bushes for spider webs.

From learning about DNA as a kid, universities are beginning to include genomics in medical school. Stanford University is one of the schools pioneering the concept.[47] How better to learn about DNA than to study your own genomic profile? The provocative Stanford class 'Genomics 210: Genomics and Personalized Medicine', and others like it, are teaching genomics as an essential part of modern medicine.

Stuart Kim, founder of the Stanford course, believes there is no better way than self-study for future doctors to learn how to incorporate this type of knowledge into the treatment of patients. Understanding our DNA can provide explanations of why we are the way we are, drug sensitivities, and future diseases we might develop. Society is increasingly aware of how genomic information can be used to make medical decisions. High-profile cases are spreading this message far and wide. One of the most famous involves the movie star

Angelina Jolie. After being genotyped, the actress opted for a double mastectomy to reduce the risk of hereditary breast cancer.[48] Future doctors will need to field questions from their patients about their genetic health. To respond effectively, physicians will need a solid grounding in genomic medicine but also an insight into how their patients might react to test results.

After students in the Stanford course use a cotton swab to get a specimen from inside their cheek, they mail it to the company 23andMe for processing. Stanford is careful about managing emotional expectations about the results that will be returned. They can be unsettling. Testing is confidential and voluntary. Students must attend informed consent sessions so they understand the types of news they might hear. Those taking part have to agree that they are willing to accept any emotional angst that comes with the findings. Students are provided access to genetic counselling and psychiatric care after receiving their results in case they want to discuss the outcomes further. If students prefer not to study their own genes, they can use a public reference sequence instead.

Students ask the questions of 'Who am I?' and 'Where do I come from?' Through DNA genotyping they gain access to information about ancestry, including which historical human population they came from—maternal and paternal lineages of humans have been mapped and dated—and how much Neanderthal DNA they harbour. Some learn things they might not have wanted to know. Predictably, at least one student discovered his dad was not his biological father. Others find increased odds of developing diseases, including a range of cancers. Understanding what 'increased odds' really means is critical. Much uncertainty of outcome remains for the vast majority of genetic variants. Along with genomics-savvy doctors, the need will clearly arise for genomic counsellors.

Advocates of personalized medicine, like Eric Topol, who has written a provocative book on the subject, entitled the *Creative Destruction of Medicine: How the Digital Revolution Will Create Better Health Care,*[49] sharply criticize the medical establishment for failing to train doctors in genomics. Courses like the one at Stanford are a start. Hopefully, the trickle of such courses will swell in the next decade to match the promise of a new era in medicine: one of medical genomics.

Genomics Goliath

Genomics is moving into the mainstream. Each year, the Massachusetts Institute of Technology (MIT) publishes its list of the 50 most disruptive companies. The businesses are defined as ones that help create a new market by displacing an earlier technology. Such innovations improve a product or service in ways that the market does not expect. Familiar names such as Google, Intel, IBM, and Microsoft abound. Only 15 of the top 50 companies listed in 2013 were on the list the previous year. The pace of technological change is brutal.

Sequencing is now big business. Google made the cut for 'running the most widely used smartphone software, which has greatly expanded the competition for devices'. Google vitals: Founded in 1997, headquartered in Mountain View California, and with 53,861 employees and a market capitalization of US$260 billion. The Beijing Genomics Institute (BGI) bounced into the MIT list in 2013 with the following attributes:[50] 'sequencing more genomes than anyone else and becoming a worldwide provider of genome services'. It was founded in 1999, headquartered in Shenzhen, China, and with 4,000 employees. Small but vast for a genomics enterprise, today the BGI has more than 5,000 employees and is the world's single largest genomics enterprise.

Taking up space in a retrofitted shoe factory, the BGI now has nodes around the world. It is a company with non-profit and for-profit arms, a mixed model that supports raising funds and doing basic as well as more applied research. BGI's president, Wang Jian, co-founded the company with Yang Huanming. Early on they managed to persuade the leaders of the Human Genome Project, then in full swing, to let them handle 1 per cent of the work. It made China the only emerging nation to play a major role in that effort. In another bold move, they secured a US$1.58 billion line of credit from China Development Bank to buy 128 state-of-the-art Illumina sequencing machines. Hot off the assembly line, they were like racing cars when most of the competition still had a regular old sedan.

In one fell swoop BGI enabled China to surpass the US in sequencing capacity. It should be noted, however, that Illumina is an American company—one named the smartest company of 2014 by MIT.[51] BGI was able to produce 10 to 20 per cent of all DNA data globally. By 2013, the BGI had 1,000 people working in its bioinformatics division alone. The average age of its employees was just 27. It did its home nation proud by leading the consortia to sequence rice and panda genomes.

The BGI claims to have completely sequenced some 50,000 human genomes by 2013, more than any other group. Its numbers are growing and involve international collaborations. To understand the genomics of weight, for example, BGI is working with researchers in Denmark to decode the genomes of 3,000 obese and 3,000 lean people. To investigate the genomic basis of intelligence, BGI is working with Richard Plomin at King's College Institute of Psychiatry in London to sequence his collection of 2,000 people with IQ scores of at least 160—that is, four standard deviations above the mean.

It comes as little surprise that the BGI holds the record for the most audacious genomics project announcement to date, the 'Million Human Genomes Project'.[52] If that weren't enough, they also threw in 1 million plant and animal genomes and 1 million microbial genomes for good measure. Little information has emerged on progress towards these goals since its announcement, however, so time will tell if the level of ambition is currently achievable. It is estimated that fewer than 100,000 people have had their whole genome sequenced at the time of writing, but some expect we will have several million sequenced by 2020.[53] These figures appear plausible, perhaps even conservative. Certainly, they are still tiny compared to Church and Gilbert's dream of 'six billion base pairs for six billion people'.

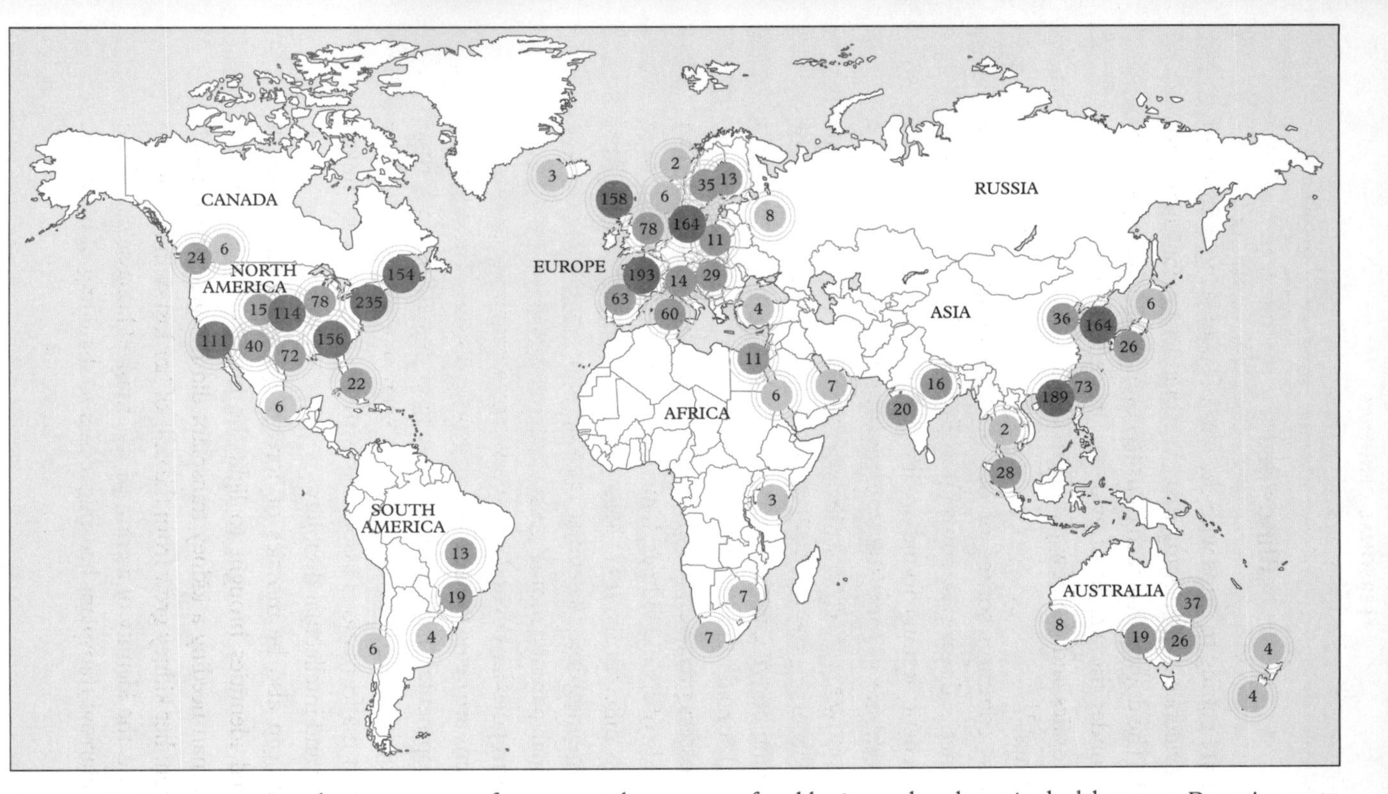

Figure 2. DNA sequencing machines are pieces of equipment that now comfortably sit on a bench top in the laboratory. Dropping costs and growing interest means sequencing centres are popping up around the world. The Omicsmaps.com site now keeps track of global sequencing capacity. This map shows the number of high-throughput DNA sequencing machines at different locations around the world. This map symbolizes the growing democratization of the world's sequencing power and the rise of the BGI in China. Talk is of a future where every laboratory has a DNA sequencer, especially as efforts to miniaturize the process become a reality.

Blindsided

Just as we are getting to grips with the idea of sequencing millions of genomes, evidence is suggesting that even one per person might not be enough. The dogma that each of us has one genome to sequence is crumbling under the weight of evidence. It seems that we might be genomic mosaics and the new paradigm could be 'one human, multiple genomes'.[54]

The most common source of our multiple human genomes is cancer. Genetic disease is conventionally thought to arise from inherited genetic lesions found in the germ line—the sperm and eggs that combine to form the first human cells from which we all grow. In contrast, cancer is a disease that can arise from genetic mutations occurring within cells in the body—somatic cells (for *soma*, meaning body). Cancerous cells are aggressive in their attempts to grow and spread to places they are not meant without permission.

We all possess precancerous or slow-growing cancerous cells. In an autopsy study of six individuals, high rates of cellular mosaicism were found across different tissues.[55] Mosaics were classified as having one or more large insertions, deletions, or duplications of DNA compared to the original 'parent genome' created at conception.

Mosaicism goes far beyond cancer. An increasing number of somatic mutations are being linked to other genetic diseases. These include neurodevelopmental diseases that can arise in prenatal brain formation and cause recognizable symptoms even when present at low levels. Brain malformations associated with these changes are linked to epilepsy and intellectual disability.

Humans can also be mosaics of 'foreign' genomes. Rare cases of confounded identities brought to light the first examples.[56] In one case, a woman needing a kidney transplant did not genetically match her children; her kidney grew from the cells of her lost twin brother. In another case, the identity of a criminal was masked because cells from his bone marrow transplant had migrated into the lining of his cheek.

Cheek swabs were taken for his DNA test. Even more remarkable, observations suggest that many women who have been pregnant might be genomic chimeras. In samples from brain autopsies of 59 women, for example, 63 per cent of neurons contained Y chromosomes originating from their male offspring (actually from the fathers).[57]

Doctors and geneticists are just starting to explore what having a multiplicity of genomes means for human health. At this point they are busy mapping the extent of the phenomenon but the message is already loud and clear: genomics continues to astonish us and genomic diversity is appearing everywhere we imagine to look, including inside our own bodies.

Beyond genomics, epigenomics is perhaps an even higher mountain of diversity to scale.[58] Genomes might be relatively static entities at the level of their nucleotides A, C, G, and T, but the double helix can be decorated in numerous ways that change how genes are turned on and off, and in which combinations. In essence, exactly the same genome sequence can have very different effects depending on its history and context. Gene expression patterns can change frequently, and in some cases the modifications are even passed on to the next generation. It never ends. Human genetic variation continues to blindside us with its enormity and complexity.

3

HOMO EVOLUTIS

BabySeq

Reading our DNA is a new way of interpreting ourselves. Current trends suggest that one day, perhaps much sooner than we think, we will all have our own genomes sequenced. We may opt in for health reasons, curiosity about our ancestry, or to push the intellectual and emotional boundaries of what it means to be human. Some of us will just want to be part of the growing 'genomics club'. It makes for interesting conversation around the water-cooler—or these days, the espresso machine.

There is increasing talk of sending all babies home from the hospital. This is just one step towards the use of genome information as a routine part of health care. How might babies with known genome sequences fare compared to those born without a crystal ball in their cradle? This is what Robert Green and Alan Beggs of Harvard University hope to find out in their BabySeq project, one of a new breed of genomics studies designed to understand the implications of our ability to read the language of life.[59]

Reading a genome sequence is like consulting an oracle in many ways—the messages must be interpreted carefully and often in the absence of a clear, or even complete, message. What actions should we take based on the word of the genomic oracle? How sure can we be? How much information can we currently glean? In the short term: perhaps not too much.

Some genes speak loud and clear, most don't—certainly not yet given our limited knowledge of the language of DNA. The classic textbook example of a very clear message is encoded in the gene that causes the disease sickle cell anaemia. Sickle cell anaemia is caused by a single amino acid mutation in the haemoglobin gene found on chromosome 1. This causes a normally hydrophilic amino acid, glutamic acid, to be replaced with the hydrophobic amino acid, valine. The result is a sickle-shaped cell with a rigid membrane. The sickle cells are not elastic enough to allow the oxygen carrying red blood cells to flow easily through the smallest blood vessels, and this causes blockages.

There are many other genetic signatures that can have clear health implications. For instance, there is the growing pharmacogenomics database to help guide drugs better. Some 100 drugs now have a Federal Drug Administration (FDA) label recommending a genomic assessment before the drug is used.[60]

Some genes only suggest the probability of certain outcomes. Take, for example, the ApoE gene, the strongest genetic determinant of Alzheimer's disease and the gene that James Watson and Steven Pinker hid from themselves. Most genetic messages are strongly context-dependent. Predictions must be viewed through a maze of complex interactions of genes, environmental factors, random events, and historical contingencies. Many genes interact so heavily with external factors we can't yet understand the consequences of possessing them. These could range from aspects of the surrounding environment, such as where you live and how much pollution you encounter, to your behaviour, such as whether you smoke or not.

If we don't like the prophecy, how might we cope? The hope of genomic knowledge is that it will improve our lives and this includes not just mitigating any genetic shortcomings, but perhaps outright fixing them. Gene therapy is an active area of research for changing nature's blueprint and chromosomal therapies are also being explored, for example, to cure Down's syndrome, a genetic abnormality that causes severe learning disabilities.[61]

The condition is caused by an extra copy of chromosome 21. In 2013, it was shown that it was possible to shut down the extra chromosome in cells from individuals with Down's syndrome grown in the lab.[62] This was accomplished by inserting the XIST gene, which usually inactivates one of the X chromosomes in women, into the unwanted chromosome. Now that we are getting more adept at reading genomes, we are beginning to recognize what underlies certain health outcomes. Like plastic surgery for the body, will we be able to sculpt our DNA one day?

Devil's Ark

There is one version of our genome we hope never to see: a sad, dismantled, disgruntled, and selfish genome. Finding a cure for human cancer is the driving force in genomics. Nowhere is more effort going into understanding the process of genomic change than in human cancers.

Each cancer has a different aetiology, or origin, and a different progression of disease and expected outcomes. There are hundreds of defined cancers and tumour types because cancer can stem from any type of cells; humans possess at least 200 different types. The International Cancer Genome Consortium (ICGC),[63] an intergovernmental collaboration, is generating a reference collection of genomes from different types of cancers, tissues, and individuals. Even this billion dollar project is only the tip of the iceberg.

Cancers are so pernicious because they achieve immortality—they don't die but acquire superhuman powers. Due to this fact, the most famous genome in history is the HeLa genome. HeLa cell cultures have been used to underpin research that has contributed to more than 60,000 scientific articles. Bought and sold, they underpin a huge industry. They helped to develop vaccines like the one for polio, cancer medications, *in vitro* fertilization, gene mapping, and cloning. A year before Yuri Gagarin became the first man to orbit the Earth in 1961, HeLa cells were sent into space by the Soviets to explore how tissues respond to zero gravity.

HeLa cells have a long and contentious history. The way the HeLa genome became immortal demonstrated the tension that can arise between personal privacy and the public benefit derived through biomedical research. HeLa cells are named after Henrietta Lacks, a young woman from Virginia with five children. After a routine biopsy of her exceptionally aggressive cervical cancer cells was taken in 1951, they were walked down the hall and put into experimental culture. They proved the first to live for any significant amount of time; when cells are healthy they quickly die.

Despite the incredible importance of her cells to science, the Lacks family were kept in the dark about the existence and use of HeLa cell cultures until 25 years after her death. They were finally told only because researchers needed family DNA to create diagnostic markers for HeLa cells lines. Scientists were worried because HeLa cells were acting inexplicably. Other independently derived laboratory cell cultures, ranging from breast and prostate cells lines, turned out to be in fact HeLa cells. HeLa cells had taken them over—they were so hardy they could float on dust particles in the air and travel on unwashed hands.

HeLa cells have a genome so different from a healthy human genome that some biologists, such as Leigh Van Valen of the University of Chicago, wanted to give them a new species name, *Helacyton gartleri*, even though they are derived directly from human cells.[64] While most scientists find this extreme and unhelpful, some researchers liken cancerous cells to microbial species: unicellular organisms on their own evolutionary trajectory.

Given how important these cells are to science it was inevitable that the genome should be deciphered. In 2013, Lars Steinmetz at the European Molecular Biology Laboratory in Heidelberg, Germany published a reference version of the HeLa genome.[65] The genome sequence provided the most detailed look to date at the genetic chaos that characterizes these cells, and to a lesser extent all cancers. HeLa genomes are chock full of extra copies of her chromosomes and riddled with errors. Large segments of DNA are shuffled, rearranging the normal order of the genes in healthy humans. Around 2,000 genes are expressed at levels higher than those of normal human tissues.

Publication of the HeLa genome proved an ethical lightning rod. Several scientists, family members, and privacy campaigners objected vehemently. The genome had been sequenced and published without getting any formal consent from the Lacks family. The fact that no consent was actually required by any of the regulatory systems in place did not assuage those who felt this was an unacceptable invasion of privacy. It was bad practice and served to highlight the need for updated regulations. Steinmetz and colleagues retracted the genome.

Rebecca Skloot, author of the best-selling book on Lacks,[66] explained why she found the genome publication morally wrong in a *New York Times* article. Recognizing the benefits of genome sequencing, she pointed out the urgent need to consider privacy issues: 'No one knows what we may someday learn about Lacks' great-grandchildren from her genome, but we know this: the view we have today of genomes is like a world map, but Google Street View is coming very soon.'[67]

A compromise was drawn up between between scientific benefit and individual privacy. In August 2013, the National Institutes of Health (NIH) unveiled the HeLa Genome Data Use Agreement.[68] It stipulates that a panel, including two representatives from the Lacks family, will review applications to access this genomic data and set conditions for its use. Francis Collins, head of the NIH and leader of the public Human Genome Project, made the announcement and stated: 'We should all count Henrietta Lacks and her family as among the greatest philanthropists of our time when you consider how they contributed to the advancement of science and our health.' HeLa cells may add to their tremendous list of accolades credit for helping to create a policy to protect genomic privacy.

Another cancer is even more extraordinary. Looney Tunes cartoons made famous an over-muscled Tasmanian Devil who spins in a tornado of fury uttering unintelligible noises and trying to eat everything in sight—most notably Bugs Bunny. In reality, Tasmanian devils are nocturnal and resemble smallish black dogs with very thick hair. The notorious devilish yowls and behaviours are just bluff, similar to a possum playing dead. The young are even cute, playful, and easily tamed.

The Tasmanian devil, *Sarcophilus harrisii*, suffers an infectious scourge that might actually result in its extinction—a cancer called devil facial tumour disease (DFTD). DFTD causes bloody-looking tumours on the faces of Tasmanian devils that grow so large the animals cannot feed and end up starving. The disease has a 100 per cent mortality rate. Cancerous cells are passed by bites thought to occur during feeding or mating sessions. Unlike most cancer cells, which are notoriously 'unique' in their exact genetic lesions, all DFTD tumours are identical to each other; the disease originated and spread from one female Tasmanian devil.

Actions are under way to save the Tasmanian devil genome. The state of Tasmania is also working to collect uninfected devils to stock a 'Devil's Ark'.[69] Genomics researchers are also working hard to protect the Tasmanian devil by interpreting the genome and protecting remaining genetic diversity. One of them, Webb Miller of Penn State, also happens to be co-author of one of the most famous tools in genomics, 'BLAST',[70] which compares DNA sequences. Miller is now leading a project that is helping to design a genomic conservation strategy.[71] The obvious strategy is to protect the pool of uninfected devils, but the dream would be to cure the cancer. The holy grail for Tasmanian devils and humans alike, is to understand genomes so well we can fix them.

De-extinction

On 5 July 1996 a sheep was born. Dolly became an overnight global sensation because of her incredibly unusual genome. It hadn't come from the normal fusion of sperm and egg genomes but from an adult mammary gland. Dolly was living proof that cloning was not science fiction and in honour of her origins she was named after none other than Dolly Parton. One sheep provided her egg, a second her

DNA, and a third carried her to full term. She could boast having not one but three mothers.

In principle, each genome has all the information needed to code the biochemical instructions for any of the types of cell that make up the organism. One genome, wrought by the union of sperm and egg, creates all the specialized parts of our bodies from kidneys to eyes, from heart to skin. Each fertilized egg starts omnipotent, but repeatedly divides to create pluripotent stem cells, and then daughter cells with specific functions.

Dolly proved that a genome taken from a cell in the somatic tissue of an adult body could be used to recreate a new individual. Recognized with a Nobel Prize in 2012, Sir John Gurdon first demonstrated our ability to turn an adult cell back into a pluripotent stem cell—one that is capable of making any other kind of cell. Genomes can be rejuvenated.

The discovery opened the door to cloning, and perhaps even bringing lost species back to life. 'Bodily, but not genetically extinct' is parlance for a species that is a candidate for 'de-extinction'. Given that DNA holds the complete recipe for an organism, that cloning an animal is feasible, and that ancient DNA can be recovered, why not use genomics and cloning technology to revive extinct species? The list of extinct creatures is long and growing daily; about 100 species go extinct each day. Famous historical losses include the dodo, passenger pigeon, Tasmanian tiger, and moa. Since genomes can be rejuvenated, it is now theoretically possible to use genomics and cloning to revive extinct species—but should we?[72]

At the time of writing, the closest we have come to de-extinction is the case of the Spanish bucardo, a species of wild goat. A tree fell and crushed the last burcado in the early 1990s, so scientists set to work trying to clone it from cell lines taken from the last few individuals. They succeeded after many attempts to create an embryo that was carried to term in the womb of a related species, but the young burcado died. It had three lungs and other abnormalities. For a fleeting 10 minutes it was resurrected, or more clumsily, 'de-extincted'.

As the case of the burcado shows, we do not understand the magic of the developmental process. A genome might contain the full recipe for an organism, but you also need to activate the right bits of genomic software, in the right way, at the right time, and in the right place. Living organisms do this exquisitely through the cascade of cell divisions that lead from single cell—how we all start—to complex multicellular organism. Successfully mimicking that process is one of the biggest challenges for synthetic biology.

DNA synthesizers have already enabled scientists to bring some viral strains back to life. The polio virus was recreated in 2002 and the 1918 flu virus in 2005. But this sidesteps the ontological process mentioned above. Viruses, with their pared-down minimalist genomes, do not develop and they cannot even replicate without a host cell. They are not living. The real test is the *de novo* synthesis of genomes of living organisms.

George Church thinks the day might come when a human mother gives birth to a Neanderthal baby.[73] The Neanderthal genome could be synthesized and placed into a surrogate cell and, assuming *Homo sapiens* developmental pathways do not differ too much from those of Neanderthals, be carried to gestation by a willing woman. Following coverage of Church's provocative suggestion in the mass media, a flurry of women apparently stated their willingness to volunteer. There are many hurdles yet to overcome on the path to de-extinction. It seems finding female human surrogates might be surmountable, although they should give pause to think about the experience of the unfortunate burcado.

Synthia

Craig Venter wants to create brand new genomes, not just resurrect extinct ones. He wants to make designer ones, biotechnologically savvy ones capable of helping with every aspect of modern life. In a 2012 interview with the *New York Times*,[74] Venter announced he was working on the 'Hail Mary Genome' project. He tasked his team to create a new single-celled organism. He planned to assemble and trial two genomic designs. If either worked it would be the first man-made genome of a free-living creature or what he called 'the first rationally designed genome'. Just in case that didn't attract enough attention, he added: 'if there were a God, this is how he would have done it'.

Our ability to make increasingly sophisticated designer bugs promises rich rewards. Many bacteria are easily grown, or cultured, in the lab in huge numbers. From the most basic chassis, novel organisms with better combinations of economically important genes could be rolled off a genomic assembly line. Long-term goals include engineering helpful microbes to solve major societal challenges, such as clean sources of energy.

Venter began this journey by picking *Mycoplasma genitalium* for genome sequencing in 1995.[75] It had the smallest genome of any organism known at the time, which could be grown in the lab. It has a minimal metabolism and little genomic redundancy. It has dispensed with the normal bacterial cell wall and relies on its host, humans, to provide most of what it needs. Its genome was as close an approximation to the minimal set of genes needed to sustain life known at the time.

With the genome of this organism in hand, Venter had a template; his team then began systematically knocking out pieces of DNA to see if an even smaller genome might still be viable. In 1999, Venter's team reported they had systematically removed each of the organism's genes, one at a time, and from these experiments concluded that perhaps only three-quarters were needed for life in the laboratory.[76] The final sentence of the paper is perhaps the first reference to synthetic genomics: 'One way to identify a minimal gene set for self-replicating life would be to create and test a cassette-based artificial chromosome.'

Venter and colleagues needed the ability to write their new genome and attention turned to synthesizing long stretches of DNA. In 2003, the J. Craig Venter Institute (JCVI) succeeded in synthesizing the genome of a bacteria-hunting virus, the model phage *Φ*X174 (phi X).[77]

The team produced the 5,386 base pair genome in only 14 days. By 2008, scaling up this technology allowed Dan Gibson and colleagues at JCVI to synthesize the 582,970 base pairs of *Mycoplasma genitalium* genome, dubbed JCVI-1.0.[78]

At the same time, the JCVI team was working on genome transplantation technologies, similar to those used to clone Dolly. By 2007, Carole Lartigue's team had successfully removed the genome from a *Mycoplasma capricolum* cell and replaced it with a genome that they had extracted from a *Mycoplasma mycoides* cell.[79] She successfully booted up one bacterial genome in the cell of another species.

The formula for 'digitally parented' life was in hand. In 2010 Venter, Hamilton Smith, and Clyde Hutchison announced they had created the first synthetic life form, Synthia,[80] or JCVI-syn1.0. To Venter, Synthia is a new species, one with a computer as its parent. It was created from the 1.08 million base pair genome of *Mycoplasma mycoides*, a smaller cousin of *Mycoplasma genitalium*, and a *Mycoplasma capricolum* recipient cell.

The culmination of 15 years of research, Synthia's DNA holds human watermarks, the final step before creating 'new life'. Like the flag that Armstrong and Aldrin planted on the Moon to show they had been there, Synthia's genomic flags include the names of Venter and other scientists involved in its creation, the Web address of the genome online, and secret messages for future readers of Synthia's genome. Literary quotes sum up the spirit of the project: 'To live to err, to fall, to triumph, to recreate life out of life' (James Joyce), 'See things not as they are, but as they might be' (Robert Oppenheimer), and 'What I cannot build, I cannot understand' (Richard Feynman).

We cannot yet synthesize the cell from scratch.[81] Venter used a genomic template largely lifted from nature and still needed an existing cell in which to boot up the synthetic genome. Nevertheless, demonstrating the capacity to rewrite the entire genomic software, the operating system underpinning cellular activity, was a major breakthrough. It offers the potential to build miniature biological factories, and these synthetic organisms promise to transform a variety of sectors, from agriculture to energy and health care.

As Venter put it in a 2013 interview with the *Wall Street Journal*: 'We are a software driven species like all biology on the planet... and the key thing that we have shown is that if you change the software you change the species.'[82] The next step is to insert not watermarks but DNA that could produce novel fuels, vitamins, or enzymes, or new medicines. Perhaps, she could even gain functions required for terraforming Mars.

There are more synthetic genomes on the factory floor, including yeast. Brewer's yeast, *Saccharomyces cerevisiae*, has given humanity bread and alcohol thanks to its ability to leaven bread and ferment sugars, but it is also one of the simplest eukaryotes in terms of genome size and numbers of genes. Long a laboratory workhorse, it is one of the best-studied organisms on Earth.

Saccharomyces cerevisiae has a genome size of 13.1 million bases, 16 chromosomes, and more than 6,000 genes, a third of which are also found in 'human form' in our genomes. Synthesizing such long strings of error-free DNA is still challenging but researchers in Britain, the US, China, and India aim to synthesize the genome by 2017 and to boot it up by 2018. Inevitably, some are calling for the Human Genome Project 2.0—a project to synthesize the human genome.[83]

Embryo genomics

You don't need to be Craig Venter to create genomes. People do it all the time, the old-fashioned way, by creating babies. Conventional reproduction still requires genetic gymnastics. The splitting of chromosome pairs in meiosis is fertile ground for error and mutation. Parental chromosomes must separate, reducing the two sets (the diploid genome) to one (the haploid genome). Doing so unmasks dud genes that were previously hidden by the second 'good' copy.

They must then marry up again in a careful negotiation. Embryos with severe abnormalities fail to negotiate the earliest rounds of cell division or end up as miscarriages. Rates of genetic lesions climb with the age of the parents, in particular the mother. They are even higher for embryos that are 'given help' beyond that which nature provides, as occurs in IVF treatments.

Connor Levy was born on 18 May 2013 in Philadelphia to a couple that had cells from their IVF embryos pre-checked for genetic abnormalities. Of the 14 embryos tested by Dagan Wells's[84] team at the University of Oxford, Connor's was the lucky one. While few would question the ability to circumvent failed pregnancies, this technology opens the door to whole genome sequencing soon after conception. Should the selection of embryos based on genomic features be considered acceptable?

In the case of severe genetic disease, the answer seems clear-cut to many. But what about picking an embryo for eye colour, hair colour, sex—or intelligence or natural sports ability? One can start to imagine potentially scary perversions that access to this type of knowledge might bring. Clearly, this technology—like most others—must be used judiciously or it risks taking a dark turn. The moral conflict at the interface of self-determination and genetic fate is a rich source for literary intrigue. It can make for popular movies too, as the 1997 film *Gattaca* demonstrated.

Dagan Wells says that the prospect of designer babies is remote, even if it were made legal. IVF produces only a dozen or so embryos at best, so the odds that one has all the traits a couple desires are very low. Would anyone go to the trouble? Given the chance, some surely would—and most likely will.

Maybe the best way to avoid the moral dilemma of having to select between embryos is to select an optimal partner in the first place. Prescreening couples for genetic compatibility before making the decision to conceive is just an extension of current genetic-counselling approaches. Usually this is only applied when two parents are both known to be at high risk of carrying alleles for a specific and serious genetic disease, like Tay-Sachs. But why stop there?

Should women gaze into the genomic crystal ball to pick the best father for their children? Lee Silverman has launched the company GenePeeks to help women looking for sperm donors;[85] 23andMe has won a patent called 'Gamete donor selection based on genetic calculations' for helping prospective parents pick the traits of their offspring, from preferred eye colour to lack of disease predispositions. It appears to be the blueprint for a designer-baby-creation system though the company denies it.[86]

Do we owe our children good genes if we can provide them? Might a lawyer one day argue that deliberately *not* giving them the best genes available is a form of abuse? It is not inconceivable to imagine a world where natural reproduction would seem primitive and even barbaric. It might even become compulsory in some countries as offspring from natural births start suing their parents for negligence. What if preselecting embryos based on their genomes—and even mates—becomes the norm and no one can remember a time when not doing this was normal?

Genomematch.com

DNA is the debut album of British girl band Little Mix, and its title track, released in November 2012, peaked at number 3 in the UK music charts. Its lyrics abound with references to X and Y chromosomes, genetics, biology, and DNA. Songwriters Jade Thirlwall and Perrie Edwards explained the lyrics as just a piece of clever wordsmithing to come up with something unique. They just started matching science words with love. If you get rid of everything scientific, they say, it's just a love song.

Their song hits the nail on the head. We use mental calculus to choose the object of our love. We glean clues from health and behaviour, smell and looks, to cultural background. We weigh up the information and assign an attractiveness rating that can change as more information is acquired. We are looking at genes, or at least for evidence of good ones.

Matchmaking is an old industry and the demand will never diminish. How much could genetics underlie the complex and magical phenomenon of love? Modern services may well exploit genomics. Companies like scientificmatch.com, GenePartner.com, and sense2-love.com have attempted to use genetics to predict love matches. They offer DNA tests that use the science behind the famous 'T-shirt smell test'; women prefer the smell of men with more different immune genes. Testing potential dating partners offers a new way to combine traditional measures of preference with newly emerging genetic ones, one where no one is obliged to smell a T-shirt.

We are starting to learn how genes contribute to our emotional health and perhaps even to how we might fare in our love lives. Robert Levenson of the University of California Berkeley led one of the first studies linking genetics, emotions, and marital satisfaction.[87] We all have two copies of 5-HTTLPR, the not-so-catchy name for the gene that regulates serotonin. The gene comes in two forms—alleles—that differ in length. The polymorphism appears to correlate with aspects of marital fulfilment.

Spouses with short 5-HTTLPR alleles were unhappier in their marriages when things were worse, and happier when things were better, compared to spouses that had one or two long alleles. In other words, they were more extreme and views of marital fulfilment matched immediate emotional conditions in the marriage. The study's authors stress that spouses with different 5-HTTLPR variants can still be highly compatible. Those with short variants just seem to be more sensitive to the emotional climate of the marriage than those with longer alleles.

Both types of variants can have advantages and disadvantages, the authors of the study stress. Their works shows that genetic differences influencing marital harmony might well exist, and if so, people will inevitably be influenced by them. The path is not likely to be easy. Understanding the genetic aspects of emotions, complex behaviours, preferences, compatibility, and love is a massive challenge.

The genomics of love is a wide-open field and certainly many would be delighted to find proof that genomic matchmaking could work. Are there combinations of genes common in people that romantically bond long term versus pairs of people drawn at random? One could certainly imagine trawling a genomic database linked to a social network to look for associations between genes in successful and unsuccessful couples. Another approach would be to design experiments that use genomic comparison to see if preferences could be predicted. Results could be compared to outcomes when people rate each other in person for attractiveness and compatibility using traditional cues.

Once more of the science is in place, might a genomematch.com await those looking for love? Dating services collect a range of information about participating singles and allow subscribers to browse for appealing matches. Some sites go further and suggest matches. A genomic dating site could work the same way. Have your genome sequenced and start to trawl for matches. Traditional and genetic information can be combined to give an even more effective search profile. Do you find lactose intolerance, baldness, bad caffeine metabolism, leanings towards addiction, or propensity for weight gain unattractive? Click 'No. Thanks.'

Certainly, the basis of attraction and love is one of the most magical things of our species. It defies logic and could never be cracked by number crunching alone—or could it? If these types of DNA-based approaches bore fruit, a genomematch.com service could be technologically easy to create. Would it ever be socially acceptable, though? Is the concept more like the dystopian movie world of *Gattaca* or a future 'genomic love' paradise? Will genomic matchmaking dampen the fires of romance or open the door to perfect love?

Humanity rebooted

Now that we can write DNA, might humans someday evolve into a new species? Usually the process of speciation in vertebrates is on the order of a few million years. Might we speed up this process and drive our own evolution? Could we one day change the human behaviour or our genome enough to create *Homo somethingelse-is*? Some think we can—and will. Juan Enriquez is one such visionary espousing radical transformation of the human species in the near future. A former Harvard pundit and 'genomic futurist', he proselytizes that we will soon bring about the 'biggest reboot of the human civilization'.[88]

Homo evolutis, he says, will emerge from *Homo sapiens*—once we gain direct control over our own evolution trajectory. He predicts our grandchildren could be a long-lived, genetically enhanced, next species of human. There have already been a posited 25 species of humans, why not another one? The technologies that will bring about this evolutionary leap are a powerful combination of genomics, manipulation of living tissues, and robotics. Google Glass—and the ability to telekinetically take photos—might be early signs of just how possible this could be.[89] Bionic limbs, the ability to replace or regenerate parts from modified tissues, like stem cells, are on the horizon. Our brains could merge with faster and much smarter computers boosting intelligence and through the World Wide Web we could absorb the planet's collective intelligence. What such superhuman would want to breed with a feeble human specimen like any of us?

Dmitry Itskov goes a step further in calling for the evolution of a new human species; he wants life extension and eventually genomic immortality. The Russian business tycoon takes the concept of radical human evolution through hybridization with technology a step further and he can shore up his vision with his wealth that runs into the billions. His idea is to create a spiritually lofty new human species by transporting our human 'essence' into artificial bodies. His roadmap, based on the five core technologies of nanotechnology, biotechnology, information technology, robotics, and genomics, aims to lead us to Utopia.

Itskov has launched his 2045 Initiative to build human avatars.[90] He aims to have shape-shifting bodies for us to inhabit, built from nanomaterials. If you ever wanted to live as the flying man, or the underwater dolphin, this is your chance. Extraordinarily, you might be able to transform between the two. You would exist outside in your human body, but control a superior, avatar body. Eventually he wants to find a way to transfer our brains, with our personalities, feelings, and memories, to digital storage devices. Once there, humans will live on in computers as Immortals.

We will, he believes, be holograms by 2045. Life is defined by its ability to replicate. If humans are immortal in this vision, will we still replicate? Or, would these machines replicate? If so, will DNA-based information have evolved to its next phase—from chemically based information in the A's, C's, G's, and T's of the DNA molecule to the 0 and 1 of computer-based information. Our evolutionary future is still to be written. Increasingly, we might be the authors at the keyboard, typing out the story of our genomic futures.

4

ZOO IN MY SEQUENCER

Elvis lives

The Human Genome Project was the first official 'Big Science' project in biology. At a cost of US$3 billion it ranks in the annals of scientific history with projects like the Moon Landing, the International Space Station, and the Large Hadron Collider. Yet we sequenced many genomes before tackling those most famous 3 billion base pairs. Walking comes before running.

Genomics started, logically enough, with the smallest genomes. The first organism to have its genome sequenced, in 1972, was the phage *Φ*X174 (phi X)—the virus of bacteria—the very same that Venter chose to be the first to synthesize *de novo*. The effort was led by Fred Sanger at the University of Cambridge, who went on to win two Nobel Prizes for his efforts.

Although many do not consider viruses to be living organisms, sequencing approximately 5,375 nucleotide base pairs was an important technological breakthrough—the first in what would be a long legacy of genomic achievements. By 1981, Sanger's team had published the 16,500 nucleotide base pair sequence of the human mitochondrial genome.[91] This tiny genome, from an ancient bacterial symbiont which now supplies the energy in our cells, has only 37 genes, but paved the way for much more ambitious projects.

In 1986 Thomas H. Roderick, a geneticist from the Jackson Laboratory in Bar Harbor, Maine, took part in an international meeting in Bethesda, Maryland about the feasibility of mapping the human genome. One evening, he and others were discussing a new journal to support the science. Wondering what to call the journal, Roderick recalls the moment he coined the term that would encapsulate a revolution in biology: 'We were into our second or third pitcher, when I proposed the word "genomics".'

To Roderick, however, 'genomics' was more than just a journal title; he saw it as 'a new way of thinking about biology'.[92] Also in the inaugural editorial, Victor Kusick and Frank Ruddle described the new discipline as one 'born from the marriage of cell and molecular biology with classical genetics and ... fostered by computational science'. They recognized the rallying call for the complete sequencing of the human genome, and suggested that it was now 'feasible or at least conceivable'.

By 1995, genomics took off in reality with the sequencing of the first bacterial genome. Craig Venter's laboratory, The Institute for Genomic Research (TIGR), completed the first genome of a free-living species, *Haemophilus* influenza.[93] As its name suggests, this bug was classified back in 1892 when it was thought to be responsible for human influenza. It was blamed for many deaths including, most notoriously, the 50 million killed during the 'Spanish flu' pandemic after the First World War. The true culprit was identified in the 1930s as a group of viruses with single stranded RNA genomes in the family Orthomyxoviridae.

Far from killing millions, we now know that *H. influenza* generally lives in peace with its human hosts. Although it might not cause flu, *H. influenza* is not altogether harmless. It can cause infant ear infections and occasionally even deadly meningitis. Given its unfortunate experience at the hands of taxonomists[94] it is perhaps fitting that *H. influenza* became the first species to reveal its true self: its 1.8 million base pairs long DNA name.

The *H. influenza* genome project was a proving ground for the radical new approach of 'shotgun sequencing'. This method circumvented the laborious process of using polymerase chain reaction (PCR) to walk along contiguous, long pieces of DNA. A genome was sheared up into small pieces and stuck into bacteria which, when they multiply, would create many copies of each piece. These were sequenced and then computers running new 'bioinformatics' algorithms were used to piece back together an 'assembly'.

Amid the rush to publish the *H. influenza* genome, Owen White, who formerly worked with Venter at TIGR and is now the associate director of the Institute of Genome Science at the University of Maryland, led the creation of the first visual genomic map of a living organism. He admits that the mischievous side of him took over. He put his own imprint on the historic image by inserting a classic phrase.

Mojo Nixon had just released a satirical song 'Elvis is everywhere'[95] and it stuck in White's head. If one looks, in 0.5 size font, it reads 'Elvis lives' on gene 'HI1127', the 1127th gene along the circular chromosome. The amazing fact is that we share 31 core genes with this bacterium—and all of the rest of life,[96] perhaps the most solid piece of DNA evidence for Darwin's theory of evolution—and given how similar all humans are to each other, we *all* have in us *most* of Elvis.

Genomic GOLD

Genomes were scientific gold and the gold rush started in earnest almost immediately. Within the first handful of genomes we filled in the trinity (Bacteria, Archaea, and Eukaryotes) from the Three Domains of Life. The second genome project, *Mycoplasma genitalium*, also by Venter, helped lead to the creation of Synthia. The third full genome to be sequenced, again by Venter's TIGR, was *Methanococcus jannaschii*,[97] with a DNA name of 1.66 million base pairs. This genome provided a smoking gun, demonstrating that Carl Woese, who discovered the Third Domain of Life, Archaea, was right: Archaea are fundamentally different in genetic make-up to Bacteria and Eukaryotes. The first eukaryotic genome came in 1996 with the greatest workhorse of molecular biology, the yeast *Saccharomyces cerevisiae*.[98] This single-celled organism took the record for the size of a sequenced DNA name to 12.1 million nucleotide base pairs.

Nikos Kyrpides, a genomics leader at the Department of Energy's (DOE) Joint Genome Institute in Walnut Creek just outside Berkeley, California, then a postdoctoral researcher in Woese's lab, was fascinated by the appearance of the first wave of genome sequences. He started to keep track of them in a spreadsheet, but the list grew so quickly he decided to expand his personal project to keep pace. His list blossomed into the Genomes Online Database (GOLD).[99] We have gone from two genome publications in 1995 to hundreds per year and a finished roster that numbers in the thousands (see Figure 3).

'Reading' the sequence of a genome, once it is in digital form, follows a usual path. The first step is to find the genes—the sequences of DNA that code for proteins. Then the genes are 'annotated' with their likely functions, largely determined by comparison of the sequences to databases of other genes with known functions. Even in the best characterized genomes, including human, there are still genes without known functions. In many organisms the number of mystery genes is longer than the list of those we think we understand.

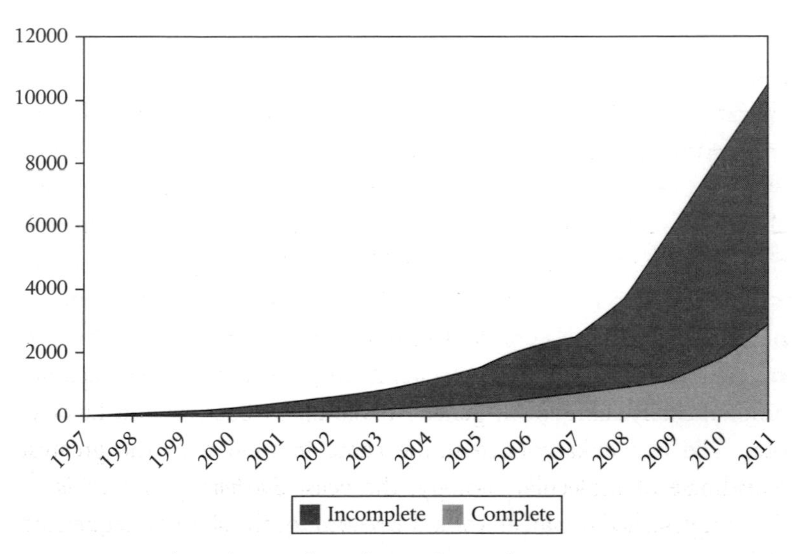

Figure 3. Growth in the number of complete and ongoing genome projects. *Source*: Genomes Online Database (GOLD).

Interactions of genes with known functions, through the proteins for which they code, are next knitted together to reveal metabolism. Finally, as other genomic data allows, functional pathways from genotype to phenotype are compared among individuals and species to identify differences and similarities that help to explain fundamental aspects of physiology and evolutionary biology.

Venter used sequencing the *H. influenza* genome and many subsequent genomes of increasing size and complexity as a way to ramp up towards the ultimate prize, the human genome. The human genome of 3.2 billion bases was sequenced in 2001, just five years after the first microbial genomes. It followed the first multicellular animal genome, the nematode *Caenorhabditis elegans* with 1 billion base pairs, completed in 1998,[100] and the first plant genome, *Arabidopsis thaliana*, with 157 million base pairs, completed in 2000.[101] We can now sequence even the largest genomes and sometimes they come from some surprising places.

The first vertebrate genome to be completely sequenced was our own. We rushed straight for the finish line without passing through the usual intermediates of biomedical research—the mouse, rat, and other primates. Venter famously told Francis Collins, the US head of the public consortium sequencing the human genome, to leave humans to Venter's privately funded effort; the public consortium, Venter offered, could 'do mouse'.

Mouse was not even the second vertebrate species sequenced. That honour went to 'Fugu', a tropical fish and sushi delicacy with a poison toxic enough to kill if not prepared expertly. In 1993, Sydney Brenner and his colleagues estimated that the Fugu genome was barely one-eighth the size of the human genome. Genome sequencing confirmed that Fugu had a mere 390 million base pairs but a similar repertoire of genes to humans.[102] It lacked difficult-to-sequence-and-assemble repetitive DNA. With little non-coding junk DNA littering its genes are signposted. Since Fugu holds many genes shared by all vertebrates, it represents a Rosetta Stone for finding genes within the animal kingdom.

Following Fugu, calls were made to sequence many more species to help understand the human genome. One of the biggest lessons learned from the early days of genomics was the power of (and need for) comparison. From the large number of mysteries in the human genome project it became clear that annotating the genomic landscape with accurate gene predictions and functional information would require more genomes.

Does size matter?

One overarching question in genomics is how genome size and complexity relates. Biochemical studies showing that the amount of DNA present in cells was not a good guide to the complexity of a species had long puzzled biologists. The sequencing of genomes across the Tree of Life has revealed some basic rules and—this being biology—some fascinating exceptions to those rules.

The prizes for the smallest genomes go to viruses. The genomes of viruses such as HIV are only some 10,000 nucleotides long. The smallest RNA (ribonucleic acid) viruses are only 300 or so base pairs. Until a few years ago it was thought that the largest viruses, like those that can cause herpes and chickenpox, were around 300,000 base pairs.

If 'who can live with the smallest genome' were reality TV, viruses would be voted off. They are not living. They cheat by exploiting a host cell to do most of the tasks we associate with life, including the most fundamental activity of all, replication.[103] It is no surprise then that viruses survive with a far smaller set of genes than 'life'; they get by with just enough to commandeer their host cell's molecular machinery.

The discovery of giant viruses rewrote the textbooks. The largest are dubbed Pandoraviruses, and have been placed into the genus *Pandoravirus* because they raised many questions about the origins of viruses and the potential existence of a fourth domain of life.[104] Pandoraviruses are up to 1 micrometre long and 0.5 micrometres across. These monster viruses are so large they were mistaken for bacteria. Moreover, they have genomes between 1.9 million and 2.5 million bases. These unusual genomes also attracted interest for the fact that only 7 per cent of their genes match anything already found in public databases. They are now being found in amoeba from around the globe. What they are doing and how they are doing it is still almost a complete mystery.

Bacterial genomes are larger, ranging up to 10 million base pairs. Archaea have a similar but narrower genome size range than Bacteria, generally not exceeding about 5 million base pairs. The smallest bacterial genomes are from endosymbiotic species. In the closest of relationships living in host cells, endosymbionts give up many of their own metabolic functions. In return, they supply the host with something it needs, such as essential nutrients that might be lacking in its diet. In the most extreme case, our own mitochondria have devolved down to only 37 genes.

Another classic example of a genome caught in the ecological and evolutionary act of undergoing genomic degeneration is that of *Mycobacterium leprae,* the bacterial pathogen that causes leprosy. The leprosy genome has only 1,600 protein-coding genes while its close relative, *Mycobacterium tuberculosis,* which causes tuberculosis, has about 4,000. Furthermore, 50 per cent of the *M. leprae* genes are dying, riddled with errors that render them 'off' and will be discarded in time. The claim for the smallest bacterial genome at the time of writing is held by the endosymbiont *Candidatus Nasuia deltocephalinicola,* which has just 112,000 base pairs and only 137 protein-coding genes.[105] This species is found in the sap-feeding insect *Macrosteles quadrilineatus,* an agricultural pest.

Eukaryotes have yet bigger genomes kept in nuclei, sacs in the middle of each cell, organized into sets of chromosomes. Animal genomes range in size more than 3,300-fold, and in land plants genome size differs by up to a factor of 1,000. The genomes of protists have been reported to vary more than 300,000-fold in size.

The largest changes occur as a result of genome duplications, or polyploidy events. *Amborella trichopoda*, a primitive understory shrub only found on the ancient Pacific island of New Caledonia, has been found to be the sole sister of all living flowering plants, the angiosperms.[106] The results of the study suggest that a polyploidy event 160 million years ago produced the ancestor of all angiosperms. We owe the existence of all flowers to the extraordinary evolutionary genomic event.

The Norway spruce, *Picea abies*,[107] is the first gymnosperm—another major branch of plant life—to be sequenced. It has a massive 20 billion base pairs (20-gigabases). The conifer's genomic corpulence is due to the spread of repetitive DNA. Transposable elements or 'jumping genes' were discovered in 1948 by Nobel Laureate Barbara McClintock and are thought to be selfish genetic sequences that can replicate and reinsert themselves into different parts of a genome; they can be seen as molecular parasites similar to viruses.

The race is on for whether animals or plants will deliver the largest sequenced genome. The marbled lungfish, *Protopterus aethiopicus*, with around 133 billion bases has the biggest estimated animal genome. Even larger, however, might be the plant *Paris japonica* with an estimated genome size around 150 billion bases. Ilia Leitch and colleagues at Kew Gardens argue in their 2010 review of genome size diversity that we need to add to our knowledge of these largest and hardest-to-sequence genomes—the 'truly obese' genomes—if 'we are to get a holistic view of genome size diversity across eukaryotes'.[108] One thing is clear: in terms of its size, the human genome is nothing exceptional.

Don't call it junk

How do we account, then, for our human uniqueness? If humans don't have an exceptionally large genome, then perhaps we have more genes packed into it than our less complex animal cousins? One of the biggest surprises of the human genome project, however, was the small number of genes found. Estimates prior to completion varied widely, but the textbooks typically quoted somewhere between 70,000 and 100,000.

The true number of genes in the human genome long flummoxed us. For fun, prior to the completion of the Human Genome Project, those responsible for finding the protein-coding sequences in the genome held a GeneSweep competition to guess the total number of genes that would be found. In 2000 expert estimates ranged from 30,000 to over 100,000, most being far too high. Lee Rowan of the Institute for Systems Biology in Seattle, Washington made a surprisingly lowball, but accurate, guess of 25,947 and won.[109]

We stand humbled. Humanity has to face the fact that its genome is not unusually big; neither in terms of its absolute size measured in nucleotide pairs, nor in terms of the number of protein-coding genes it contains. Subsequent work has, in fact, pushed the estimate down to about 20,000. Tomato plants and the water flea, *Daphnia*, both have 10,000 more genes than humans. As it happens, we have about the same number of these genes as the decidedly unglamorous worm, *C. elegans* (see Table 1).[110]

This is all part of a genomic paradox. It has long been known that genome size does not correlate with number of genes. Despite its huge genome, the conifer *P. abies* has only 28,354 protein-coding genes, much the same as the weedy model organism *Arabidopsis thaliana*,

Table 1: A comparison of the number of genes across five organisms showing the human genome has almost the same number of genes as a worm (*C. elegans*).

Species	Genome Size	Number of Genes	Year Published
Human	3BN	~20, 000	2003
C. elegans	1BN	~18,000	1998
E. coli	5MB	~5,000	1997
Baker's Yeast	13MB	~6,000	1996
H. influenzae	1.8 MB	~1800	1995

which manages with a genome some 100 times smaller. Shockingly, a microbial eukaryote responsible for urinary infections, *Trichomonas vaginalis*,[111] has just 160 million base pairs in its genome, and no mitochondria, and yet contains as many protein-coding genes as its human host.

This unintuitive relationship is historically known as the C-value paradox. Much of the variation in complexity between eukaryotic species is explained by differences in the percentage of 'non-coding' DNA (ncDNA) that their genomes contain.[112] This was a surprising result given that ncDNA was thought to be the result of genome duplications and the proliferation of repetitive elements. In other words, ncDNA did not particularly help the organism's fitness and might even harm it; in the popular press, it even became known as junk DNA.

How much of any genome is junk? Known protein-coding regions, called exons, compose less than 3 per cent of the human genome. To paraphrase Craig Venter once more, sequencing the first human genome was just the beginning of the beginning. The ENCODE (Encyclopedia of DNA Elements) Project[113] discovered that much of this non-coding DNA turns out to be transcribed, or turned into RNA. It is messenger RNA that is turned into proteins, but now we know there are many types of RNA that just stay RNA and perform a range of functions as well.

At first all this non-coding transcribed RNA was thought to be background noise, but then, as technological capacity to detect transcribed RNA in cells—the transcriptome—improved, it was evident that the background noise was different in different tissues and cells. Clearly this would not be the case if it were just random transcription of useless sections of the genome. These signals had not been detected before because scientists and their tools were focusing on the coding part, that is, the sequences that were transcribed into RNA which then built corresponding proteins.

The non-coding DNA (ncDNA) codes for things after all. Their main function seems to be the complex regulation of how proteins are formed, where, when, and in what quantity. In other words, RNA determines the proteome of a cell and means that one protein-coding gene can produce more than one protein. The ncDNA is responsible for this through complex splicing of messenger RNA during the production of proteins as well as the regulation of gene expression. That ncDNA plays such a pivotal role in regulating the proteome represents a paradigm shift in our understanding.

Complexity arises on more than one level. The number of protein-coding genes is therefore just one measure of complexity. Much more complexity can be achieved, from the same number of protein-coding genes, through the regulation of their expression patterns and differential splicing of the RNA they encode. John Mattick and colleagues pointed out in a 2007 paper that the complexity of organisms is strongly correlated with the amount of non-coding DNA; humans have a lot more ncDNA than worms for example. Indeed, the ncDNA might be the part of the code that makes humans human and not worms. Just as *H. influenza* is saddled with a misleading name, ncDNA is turning out to be a misnomer of the highest order.

Perhaps humans have more genes than we thought after all. The vast majority of the human genome is ncDNA and much of it is decidedly not junk. In fact, when coding for RNA is allowed as a definition of gene as well as coding for proteins directly, then the human genome might have over 160k genes.[114] The exploration of ncDNA is a revolution in the making and Mattick at least believes it to be the single most transformative discovery resulting from the Human Genome Project to date. One person's junk is another's treasure. The human genome was a wondrous map; it has opened up new horizons and the race is on to explore the rest of this brave new genomic world.

The first tweenome

We are now so good at sequencing we can do it in almost real time—at least this is a goal. Nowhere would this be more valuable than in addressing the cause and cure of pathogen outbreaks. Luckily, pathogen genomes of viruses and bacteria are small in size and therefore quick to sequence and analyse. This builds on 15 years of experience, much of it focused on sequencing bugs that make us ill.

Within the first ten years or so of genomics we completed the genomes of all 200 or so major bacterial pathogens of humans. As reference genomes became available for each species, attention turned to looking at the tremendous variation among genomes *within* each species. Sometimes a genome of 5 million base pairs can differ by up to 1 million bases between strains of what are the same species. This led to the important concept of the 'pan-genome',[115] the idea that there is a species level of genome of many more genes than are found in any one bacterial strain. This is also called the species 'gene pool'.

In 2003, during the SARS (severe acute respiratory syndrome) epidemic it took 19 days to obtain the first genome sequence of the virus. In the 2009 swine flu outbreak, however, within 19 days researchers had sequenced hundreds of viral genomes, determined when and where it was likely to have begun, published several scientific papers, and begun to develop a vaccine. Reviewing this advance Jennifer Gardy of the British Columbia Center for Disease Control predicted that an era of pathogen genome surveillance had arrived. Her vision of public health 2.0 revolved around regular sequencing of each city's sewage plants—a new field of 'sewage-nomics'.

In May 2011 a serious outbreak of food poisoning started in northern Germany caused by a strain of Shigatoxin-producing enteroaggregative *E. coli* (*Escherichia coli*) (EAEC) known as O104:H4. The illness caused bloody diarrhoea and in the worst cases haemolytic-uraemic syndrome (HUS), which led to kidney failure and death in 53 people.

The first draft genome of the bacteria responsible was available from BGI in China within just three days. The BGI went for release of all data; it hit the public domain as soon as it came off the sequencing machine and was announced on Twitter. Within 24 hours Nick Loman of the University of Birmingham had posted the first assembly via his blog and Twitter and it started a flood of further annotations and assemblies in response. The geographic origin of the outbreak (Germany, not Spain) was confirmed by DNA analysis and it was shown that the bug had acquired an extra piece of DNA that made it so virulent.

Justin Johnson from EdgeBio made history at the 'Copenhagenomics' 2011 conference in Denmark by posting a Tweet on Twitter post calling this event the 'Tweenome'. Genomics rode the social media wave to the end. In August the outbreak was over and BGI's *Gigascience* journal editor, Scott Edmunds, posted a blog entry summarizing the event and the unusually rapid publication of papers that ensued.[116]

Today, the Tweenome is a lasting symbol of data shared on a global level to help people.[117] A picture of the genome appeared on the cover of the influential *Science as an Open Enterprise* report from the Royal Society as an example of intelligently open data.[118] The Tweenome received the first digital object identifier, or DOI, of any BGI data set. The British Library granted it in the same manner as it does for books. Its DOI secures this DNA book a lasting future within our human-penned collections. We are now building a new wing of the universal library, one of the DNA genre.

Denisovan girl

Lewis and Clark had a very special biological mission in addition to creating the first map of the Western United States. Jefferson instructed them to search for mastodons and animals known only from fossils. American mastodons disappeared more than 10,000 years ago with the rest of the Pleistocene megafauna including super-sized tree sloths, sabre-toothed tigers, and 9-foot-long sabre-toothed salmon that hunted along the coast of the Pacific. Only 200 years after Lewis and Clark confirmed such animals really were extinct, we had their first genomes in the launch of the palaeogenomics era.

In 2007 the first complete mitochondrial DNA genome was obtained for a mastodon, producing the oldest mitochondrial genome sequenced to date. In 2005, 28 million base pairs of mammoth genomic DNA were decoded with new 454 sequencing technology and by 2008 most of it.[119] This was the beginning of a race to sequence the remains of ancient life—and the older, the better.

Only 1 per cent of the life that has ever lived on Earth is extant today; 99 per cent of the estimated 4 billion species that have ever lived are gone. Researchers are continually pushing back the frontiers to examine extinct life. Fossils are fair game as new sequencing technologies that target short pieces of DNA make sense of highly degraded and fragmentary genomes. The diversity of life being studied is vast and ancient DNA is being used to resolve diverse debates, from reconstructing human migration patterns to determining the species composition of Pleistocene ecosystems.

In 2010, the first ancient genome was fully sequenced—that of a 4,000-year-old palaeoeskimo. In 2012 a new record was set for the age of a genome: 80,000 years. DNA was extracted from a tiny chunk of the pinky finger bone of what was thought to be a Neanderthal—instead it was a more distant relative of humans, a Denisovan.[120] Svante Pääbo of the Max Planck Institute for Evolutionary Anthropology in Leipzig, Germany, a savant of palaeogenomics, led the Denisovan genome project, having already succeeded in sequencing the Neanderthal genome. Using a new method of amplifying DNA, Pääbo's team produced a genome of similar quality to those obtained from fresh material. The only other Denisovan remains at the time were two teeth. While there are hundreds of specimens for Neanderthals, Denisovans first came into our view by virtue of the DNA of a young girl.

At the time, no one thought preserved DNA of that quality could be obtained from a sample thought to be 50,000 years old. Counting the number of differences between the genomes of Denisovans and modern humans suggests the two lineages split between 170,000 and 700,000 years ago and that she was more closely related to Neanderthals and chimps than to us. Strikingly, this analysis also put back the date of her fossil at least 30,000 years. Not only had a new record been set but genomics had been used for the first time to date a fossil.

The record climbed to a whopping 700,000 years in 2013 when the multinational team led by Eske Willerslev of the University of Copenhagen reconstructed the genome of the Thistle Creek horse.[121] Found near Thistle Creek in the Yukon, the specimen was so well preserved it was possible to reconstruct 73 prehistoric horse proteins in addition to retrieving DNA from the hordes of microbes that covered it. Only 2 per cent of the DNA was endogenous, making the feat all the more incredible. By comparison the Denisovan DNA was extremely well preserved. It contained 70 per cent endogenous DNA making it comparable with modern bones.

Can older genomes be reliably retrieved? In the 1990s debates raged about whether DNA from fossils millions of years old was real. Many simply did not believe in principle that the chemistry of DNA was robust enough that useful sequences could be reconstructed from anything so old.

Reports were marred by worries of contamination; the study of ancient DNA deals with such minute quantities of degraded material that it is easily compromised by mixing with other sources of DNA. A famous paper appeared in a top journal claiming to have retrieved a DNA fragment from a Cretaceous period dinosaur. It turned out instead that the sequence was actually a nuclear copy of an ancient human mitochondrial gene. It was so divergent that it was confused for a while with the hypothesized sequence of a dinosaur.

In 2012, Morten Allentoft and colleagues decided to settle the issue of the half-life of DNA.[122] They did a rigorous study of the bones from moas: extinct, flightless birds, found only in New Zealand, that were wiped out by overhunting and habitat decline shortly after the first humans, the Maori, reached the Polynesian islands around AD 1400. Big bones buried everywhere offered an excellent model system; towering some 12 feet tall, moas were the dominant herbivores in New Zealand for thousands of years. Examination of 158 bones, aged from 500 to 6,000 years, revealed that the half-life of DNA is only 521 years—very short, for example, compared to the half-life of uranium-235 at 703.8 million years. Samples in New Zealand are too fragmentary to piece together after some 1.5 million years and only the coldest of regions might expect to fare any better.

Interestingly, they found that age only accounted for 38 per cent of the variation in DNA quality found in the samples. Clearly other factors that influence quality of preservation are at work. This gives researchers like Willerslev hope they will be able to find readable DNA more than a million years old. Even under the best preservation conditions on Earth, however, such as the frozen tundra, DNA is likely to be completely destroyed by 6.8 million years. So, we will only being seeing dinosaur DNA in movies like *Jurassic Park*.

Single-celled sisters

All life originated from a single-celled ancestor dubbed by scientists the Last Universal Common Ancestor, or LUCA. We don't know exactly what LUCA was but we certainly know it would have been single-celled and probably quite similar to the simplest bacteria. LUCA would almost by definition have had a very special and highly minimalist genome containing what we propose to call the 'Minimum Information for a Living Organism', or MILO. The MILO genome would have met the basic requirement for life, the ability to self-replicate. This is the magic genome that Craig Venter and his team are looking to use as the basic chassis for industrializing synthetic biology—hoping to do for the bio-economy what Henry Ford did for industrial manufacturing.

We sequence genomes to understand the evolution of life and our pre-human origins. For 2 billion years life on Earth was composed of relatively simple, single-celled organisms. Over the aeons, however, some of these microbial species learned to bond together into aggregates of various kinds. Then two prokaryotic cells came together in a deep symbiosis that formed eukaryotic organisms. With more time and new sets of molecules for self-recognition, cell–cell bonding, and genetic machinery, however, some of these single-celled eukaryotes made another massive leap, into symbiotic colonies of clonal cells. These were the Earth's first multicellular organisms, the 'metazoans'.

The human genome is young, a recent twig on the tree of life, but it was billions of years in the making. LUCA is our mother too and we have collected up genes from our historical pedigree of ancestral species. Comparison of the human genome to other organisms shows that 37 per cent of our genes come from bacteria, 28 per cent from eukaryotes, 16 per cent from animals, and 15 per cent from vertebrates. Only 6 per cent are novel and appeared during our evolutionary specialization into primates (see Figure 4).[123]

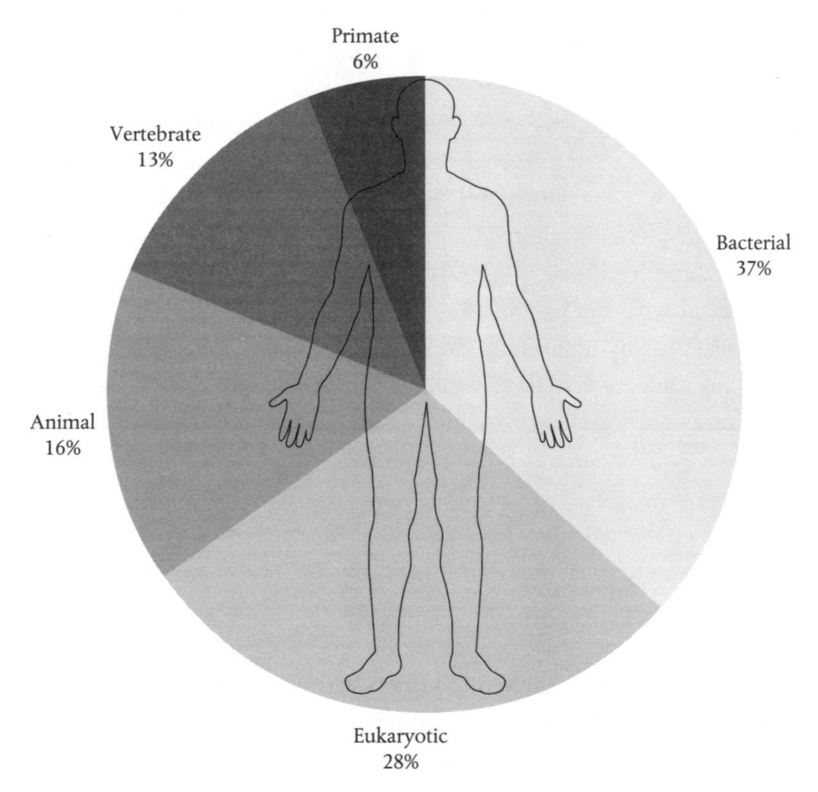

Figure 4. The origin of human genes during evolution. More than a third of our genes evolved in our bacterial ancestors and only 6 per cent in our time as primates.

For example, we now have a genome from our closest single-celled sister, the marine choanoflagellate *Monosiga brevicollis.*[124] She is a descendant of the special single-celled organism which gave rise to all the animals, including humans. This species grows easily in the laboratory in pure culture and had been the favourite subject of many phylogenetic studies of the origins of animals. With a genome size of around 41 million base pairs, it is similar in size to filamentous fungi and other free-living unicellular eukaryotes but is far smaller than the animals it gave rise to.

This small genome has 9,200 proteins, almost half as many as humans. Comparisons confirm we shared a common unicellular ancestor about 600 million years ago. Felix Dujardin, a French biologist, suggested a close relationship between animals and choanoflagellates as early as 1841, shocking because this pre-dated Charles Darwin's publication of the theory of evolution in the *Origin of Species* in 1859. Our single-celled sister feeds on bacteria and is a key component of the global food chain. She lives all over the world in marine, brackish, and freshwater environments from the Arctic to the tropics, from shallow to deep water. We are more closely related to her than our nearest relative the fungi (plants are more distantly related). You are likely quite close to one of her kind, an ancient sister, if you are anywhere near water right now.

Microbial Earth

In the history of genomics, a shift towards the systematic sequencing of genomic diversity occurred around the publication of the 1,000th microbial genome. A new breed of genome project spawned of necessity, one that aimed to fill the gaps in our knowledge of genomic diversity at the gross level. The 'Genomic Encyclopedia of Bacteria and Archaea' project, known as GEBA, was the first 'megasequencing' project to systematically explore natural genomic diversity.

The GEBA team wanted to see what genomic secrets nature still had in store. How much novel metabolism could there yet be? How many more genes to find? Most of the time, microbes had been selected for genome sequencing because of their unique physiologies. Some special trait, function, or role made them of priority interest. Now, enough 'model' genomes had been completed and the time was ripe to maximize the sheer diversity.

Nikos Kyrpides, Jonathan Eisen, Phil Hugenholtz, and Hans-Peter Klenk spearheaded the hunt for a diverse collection of microbes. They scrutinized the Tree of Life and selected strains from the 'darkest', least-represented branches. The Deutsche Sammlung von Mikroorganismen und Zellkulturen, known as the DSMZ, provided the DNAs for free. They correctly calculated that complete genomes would raise the value of their cleverly named 'buyodiversity' catalogues.

The analysis of 54 GEBA genomes appeared in 2009.[125] Sequencing of these novel genomes spiked the world's cumulative list of known proteins, showing we are still in the steep ascent of our genomic learning curve. GEBA was just the opening salvo of the much larger effort now tracked in the Microbial Earth Project (MEP). GEBA will sequence all 11,000 type strains, still a drop in the ocean of microbial diversity.

Dark matter is the invisible-matter component of the Universe that physicists know must exist but the nature of which is unknown. The similarity to the biological world is apt. Most microbes, some estimate as many as 99 per cent, have such mysterious lifestyles and nutritional needs that science cannot yet fathom them.

While the GEBA project is focusing on 'type strains' or domesticated bacteria that can be grow in culture, a special subproject focused on the truly exotic microbes.[126] The Microbial Dark Matter Project[127] targeted uncultured microbes using single-celled genome sequencing. Tanja Woyke, who led the project, focused not on the 88 per cent of all cultivated microbes that fall into just four of the 60 bacterial phyla, but on all the rest. She and her team assembled microbes from the most exotic locations, including hydrothermal vents, an underground gold mine, and bioreactors—the deepest outliers in evolutionary space.

Their exotic census tapped 29 of the deepest and most novel branches of the Tree of Life and produced 201 novel genomes. More than 20,000 new gene families emerged. Despite these DNA riches, it appears that another 16,000 carefully selected cells from the furthest corners of the bacterial tree need to be sequenced to cover just half the world's unstudied microbial lineages. At least we have a toe in the door of understanding microbial diversity.

Losing the Acropolis

Gerald Durrell pioneered the concept of the modern zoo. One of Durrell's most loved books, *A Zoo in My Luggage* (1965), recounted his year-long search for a place to call his 'own zoo'. Durrell, a prolific writer and one of the best-known naturalists of the 20th century, was a pioneer of wildlife conservation. After financing and leading expeditions to build zoo collections for others, he wanted his own zoo. He loved the animals but he also wanted to help stem the loss of species he was witnessing at the hands of man.

Durrell's first-hand experience of biodiversity loss across the world shocked him. He believed the extirpation of a species to be a criminal offence. The loss was as abhorrent as failing to safeguard anything that cannot be recreated or replaced, whether a Rembrandt or the Acropolis. The panda or the pink fairy armadillo is equally worthy. Durrell was a world-famous conservationist and part of his tireless efforts focused on reshaping the mission of modern zoos.

In 1958, he founded his zoo, the Jersey Zoological Park, in the grounds of a manor house on the Isle of Jersey. Durrell's animals had to wait a year for their zoo. To the dismay of his neighbours, Durrell's wards stayed in their crates in his backyard after returning from a six-month collecting trip to Cameroon. Durrell had a zoo in his luggage; today we have a zoo in prep for our DNA sequencers. Many, like Cholmondeley the chimpanzee and Bug-Eye the bush-baby, already have reference genomes.

Zoos, natural history museums, botanical gardens, and culture collections are stewards of genetic legacies. Pedigree analysis and determination of paternity, for example, helps reduce inbreeding in zoo animals and suggests potential mates. The study of genetic variation left in natural populations helps their management. Now conservationists are scaling up to use conservation genomics to understand and protect the world's biota. Zoos are modern arks.

One zoo leads the pack in DNA sequencing. It has launched one of the biggest genome sequencing projects envisioned to date. The world-famous San Diego Zoo is renowned for taking a pioneering role in the care, breeding, and conservation of endangered species, including the establishment of the Institute for Conservation Research. Oliver Ryder, director of genetics and Kleberg chair at the institute, has spent a career using DNA studies to aid organisms under threat, in particular through his founding of the Frozen Zoo, a collection of tissues and gametes.

In 2009, he, Steven O'Brien, and David Haussler—known as the 'Three Amigos'—launched the Genome 10K project (G10K)[128] hoping to generate genomes for 10,000 animals, one for every vertebrate genus. G10K, a deep collaboration with the world's best zoos and museums, got off to a promising start with the BGI offering to sequence the first 100 genomes on the list of 10,000.

Darwin would have been proud and not a little intrigued that one of the first genomes completed was that of the medium ground finch, *Geospiza fortis*, one of his finches from the Galapagos. To study how genetic and morphological variation underpins adaptive radiations—the rapid speciation seen on these islands as birds adapt to new niches and diverge—researchers are using the finch genome and 40 years of song recordings to explore the genetics of vocal learning.[129]

The 5,000 Insect Genome Project (iK5) has been called 'The Manhattan Project of Entomology'.[130] E. O. Wilson of Harvard, one of the most famous biologists of this century and ant expert, famously said insects are the 'little creatures who run the world'. Insects constitute approximately 53 per cent of all living species, with one group alone (the ants), accounting for almost a quarter of terrestrial animal biomass.

Insects serve as pollinators to more than 75 per cent of flowering plant species and consume or damage more than 25 per cent of all the agricultural, forestry, and livestock production in the United States, costing the economy more than US$30 billion annually. The annual cost of diseases carried by insect vectors is estimated at almost US$50 billion. Marine invertebrates, like crabs, echinoderms, corals, and molluscs, are being tackled by the Global Invertebrate Genomics Alliance, or GIGA project.[131]

Today, the Microbial Earth Project is working from the bottom up, from the smallest invisible organisms, and the G10K from the top down, from the largest, most high-profile species we know best. Filling in the middle are those working on all the rest of life and new projects are emerging.[132] The Global Genome Initiative (GGI)[133] has proposed sequencing roughly one exemplar from half of 'genus-sized' clades, which sums to about 80,000–100,000 representative genomes. Given the cost and scale of this genomic diversity project, its architect Jonathon Coddington, of the Smithsonian Institution's National Museum of Natural History, is cleverly focusing first on 'futuromics'. His team aims to lock down DNA from all these species in freezers for a time when sequencing costs drop. As international consortia assemble and divide up the genomic world and ready it for sequencing, we are seeing the lift-off of an unannounced 'Planetary Genome Project'.

5

NO ORGANISM IS AN ISLAND

The biodiversity within

As genomics turns its attention to the natural world around us, it is also revealing incredible biodiversity a lot closer to home. We are not alone in our bodies; we are not even a majority. There are 'others' among us. We are chock full of microbes. Each person contains at least 10 times more bacterial cells than human cells. We only look human because our cells are so much bigger.

We carry around about 3 pounds worth of bacteria. If it were an organ, our microbiome—all the microbes in our body—would be about the same size as our brain. Most of these bacteria are in our gastrointestinal tract, making it possible for us to digest and derive energy from our food. Despite the importance of microbes to the integrity of our existence, science has paid little attention to our 'good' bugs. Until recently, pathogens were our focus and for good reason—think of cholera, tuberculosis, or meningitis. Bacterial infections cause the deaths of more people than wars. The influenza pandemic of 1918–19 alone killed 20–40 million people, more than the death toll of the First World War. Yet, not all the microbes in our bodies cause disease. If they did, we would never survive beyond infancy.

So what is this microbiota and what are they all doing in our guts, on our skin, and in every orifice of our bodies? The completion of the Human Genome Project in 2003 opened the doors to a range of new projects, not least because there was suddenly a lot of sequencing capacity sitting around looking for something new to do. Champions led a charge for large-scale funding of systematic human microbiome studies.[134] Luminaries such as David Relman and Stanley Falkow dubbed it the 'second human genome project'.[135]

The Human Microbiome Project (HMP)[136] launched in 2007 with five years of funding from the National Institutes of Health (NIH). With US$170 million in funding it was the little sister of the Human Genome Project, but in time it would produce insights that promise to revolutionize human medicine. The Human Genome Project provided a map of human genetic parts; the HMP aimed to fill in the landscape of our microbial biota.

The HMP was sequencing not just one species but all species in these newly examined environments. The organisms of interest were distributed across a vast continent, the human body. For the first time, biomedical scientists were faced with the challenge of thinking as ecologists. Like their 18th-century counterparts in the age of exploration, microbiome explorers set out to chart a new world.

In 2006, Dr Karen E. Nelson, now president of the east coast J. Craig Venter Institute (JCVI), led the first metagenomic microbiome study of two healthy adults. Her study showed that the human microbiome, composed of 10^{13} to 10^{14} microorganisms, contains 100 times more genes than the human genome.[137]

Then, in 2010, as a leader of the HMP, Nelson led the publication of 178 reference genomes from the human gut. The combined data revealed more than half a million genes, including 29,987 that were unique to this data set. Meanwhile, across the Atlantic, a European-funded project on the 'Metagenomics of the Human Intestinal Tract' (MetaHIT) reported finding over 1,000 bacterial species in 124 Europeans sampled, with about 160 species occurring in each individual.[138] These studies revealed that even after decades of medical research, we knew remarkably little of the species that inhabit us. We knew even less about their genes—even though they outnumbered those in our own genome.

The first human microbiome studies looked at healthy, disease-free individuals aiming to establish a base line of 'normal' patterns of microbial diversity. In the HMP, more than 5,000 samples were taken and sequenced from 250 healthy volunteers in two US cities. Each subject was sampled at 15 or 18 body sites (nine oral, four skin, one nasal, one stool, three vaginal). To investigate potential changes over time, roughly half of subjects were sampled at up to two additional time-points.

Initially HMP's studies revealed that all body areas differed, not only in composition, but also in ecological organization. The oral microbiome was as diverse as the gut, but more of this diversity was shared between individuals. Common sense suggests it is easier to share mouth microbes than ones deep in your colon. Skin diversity, sampled at the inner elbows and behind the ear, was low both within and between subjects—probably because skin is exposed to a fluctuating environment. The region of the human body community with lowest diversity was that of the vagina; most women were dominated by one of four single species of *Lactobacillus* and the vaginal microbiome becomes even less diverse during pregnancy.

We can never ignore our microbes again. Or, rather, we now know that doing so would be at our own peril. The HMP and parallel international efforts formed one of the biggest coordinated mega-sequencing projects to date. They created a vast reference data set and showed we have just scraped the surface of human microbial diversity. As we learn about our microbiomes, there seems to be no end to the ways in which they impact us—for good and for ill.[139] The microbiome helps us digest our food and ward off enemies. Unbalanced microbiomes have been implicated in diseases from diabetes and atherosclerosis to asthma and autism.

We are microbial beings. Some fear the identity crisis that integrating computer chips and bionic parts into our flesh-and-blood bodies might trigger—the cyborgs of the future. First, however, we have to accept that much of our body and our genetic capacity already resides in a multitude of species other than *Homo sapiens*. Should this discovery lead us to rethink what it means to be human? Perhaps not, but we must acknowledge at least a modest shift in our sense of identity. We are biological chimeras—a 'microbial-human'.

Ratios matter

Obesity is a global problem and weight loss is a major obsession, at least in the rich world. Microbes help us derive energy from food and it has long been known that the trillions of microbes in the human gut help to break down otherwise indigestible meals. Mice studies had already shown that transplanting gut microbiota from normal animals into animals bred without the microbes increases the latter's body fat, even if there was no change in their food consumption. Microbes enable their hosts to assimilate extra calories.

In a diet-mad society, might the microbes we cultivate in our guts play a role? Might changing them provide a panacea for the obesity crisis? These burning questions have galvanized the application of genomics to human microbiomes because it seems that a strikingly simple pattern exists: the ratio of the two major groups of bacteria in our guts, Firmicutes and Bacteriodetes, appear to influence our weight.

In 2006, Ruth Ley and Jeff Gordon at Washington University found that the relative proportion of Bacteroidetes compared to Firmicutes is lower in obese people. Furthermore, obese people showed an increase in the relative abundance of Bacteroidetes following diet therapy.[140] In 2013, Gordon's group followed up with a study of identical human twins who had significant difference in weight.[141] Transplanting their microbiomes into mice showed that microbiomes from the heavier twin made the mice heavier than transplants from the lean twin.

This relationship appears to hold true at the biogeographic level as well. In 2014, a research team from University of California Berkeley and University of Arizona took advantage of the growing public database of human microbiomes from people distributed across the globe. They found a remarkable pattern that suggests our gut microbiome might be adapted to the climate in which we live.

It is well known in ecology that animal body size tends to increase at higher latitudes—a pattern known as Bergmann's rule that presumably reflects adaptation to cooler climates. Taichi Suzuki and Michael

Worobey reported that their data crunching had revealed a similar latitudinal gradient in the proportion of Firmicutes to Bacteroidetes in human gut microbiomes. People living further from the tropics tend to have more Firmicutes and a larger body mass, and this was true even when controlling for age, gender, and race.[142]

Interestingly, for one of the first times in medical history we are not talking about one or a few microbes, as in the case of most diseases, but rather of a fundamental shift in the wholesale abundances of very different types of bacteria. Such thinking represents a new paradigm for understanding microbes and our health. The early findings have triggered an explosion of interest in the human microbiome. We now know that we each harbour our own microbial fingerprint, the result of chance acquisition over time, our environment, and our genetics. We also have a direct impact on it through our eating habits and whether or not we've taken antibiotics.

Eating for trillions

There is a delicate balance to keep with our microbiomes. Many species of microbe are potentially good or bad for us; it depends on the circumstances. Their abundances matter to our health and well-being in ways we are just starting to realize. As this realization takes hold, there is increasing interest in how we should actively set about tending our microbiome. How do we take good care of the biodiversity within us?

Michael Pollan, an American author and professor of journalism at Berkeley, learned more about his own gut bacteria through the American Gut Project. As a result he wrote that he 'began to see how you might begin to shop and cook with the microbiome in mind'.[143] Pollan is an expert on fermentation—like kimchi and beer—and he is now fascinated with how fermentation might work within our bodies. Science journalist Carl Zimmer also picked up on the subject; reporting on the new field of 'medical ecology', he likens our guts to a microbial garden we must tend to make beautiful.[144]

We feed our microbiome. Just as women are eating for two during pregnancy, perhaps we should remember that we are eating for our trillions of bacteria. Modern life may actually threaten the health of our microbiome. We eat sterilized food and lead ever more sterilized lifestyles. We already know junk food is bad for us but there are many potential threats to tending a good crop of microbes. These range from pesticide residues and other pollutants to a lack of fibre and the overuse of preservatives, other food additives, and antibiotics. Could all this be driving a healthy, human microbiome out of existence? Guidance on how to keep our microbes healthy, whether by diet or consumption of particular pre- or probiotics, seems set to be a bigger business in the future.

We need to end the war on microbes. As we look at the microbial world through a new lens we are beginning to understand that our microbial biodiversity helps us in so many more ways than it hurts us.[145] The new perspective is a profound change and is influencing how we view ourselves. It moves away from the language of war to the language of ecology. It reflects a more nuanced approach that views our microbiota primarily as our oldest ally—one that provides us with health-associated ecosystem services.[146] It is another example of how genomics is disrupting our sense of identity and leading to new paradigms. We are carrying around a whole little world inside us, but where did it come from and why do we have it? What does it really do for us? What would we be like without it?

Answers are emerging. For example, microbiome research has explained the long-standing mystery of why mother's milk contains sugars that are costly to make but which their babies are unable to digest. Human milk is rich in nutrients and includes everything a rapidly growing infant needs for survival. Mother's milk seems perfectly designed for its purpose, so why would it contain energy-rich sugars that couldn't be broken down by babies? Natural selection does not generally permit much wasteful behaviour. The answer is that the sugars weren't for the baby in the first place—well not directly. They are there for the 'good' bacterium, *Bifidobacterium infantis*, named as it dominates the guts of newborns. This bacterial species is uniquely capable of harvesting the sugars in breast milk and therefore gains a competitive growth advantage over other microbes in the gut. While *Bifidobacterium infantis* happily proliferates, it crowds out less desirable bacteria. This is especially important in the naïve gut where the stable gut microflora is just being established.

Bifidobacterium infantis provides an essential biological service to the host, a special type of defence termed 'invasion resistance'. The health of our guts is defined as much by the presence of good bacteria as it is by the absence of bad bacteria. These good guys also act to nurture the integrity of the lining of the intestine, protecting babies from infection and inflammation.

Breast-feeding is nature's solution, but dried *Bifidobacterium infantis* stocks can be purchased for bottle-fed babies. At the time of writing, one 60-gram bottle of powder costs around US$50 and it is claimed that a daily intake of 1 gram should provide 4 billion viable cells. We have no expert opinion on the efficacy of this particular product, but it shows how microbiomics is reaching the marketplace. It illustrates how knowledge of our microbiome could provide a range of new options for improving our health.

Even more radical approaches are on the horizon. Might we change ourselves with the synthetic microbes we create? The US Office of Naval Research (ONR) is investigating whether the engineering of synthetic microbes could help protect against depression and obesity.[147] To that end, ONR has enlisted Rice University's Jeff Tabor, who previously conducted research that used microbial communities to perform image-processing tasks. Tabor hopes in the long run that microbes could even be engineered to act as doctors. He hopes they could be engineered to detect combinations of molecules that indicate poor health of the gut. Once detected, the microbes would respond by turning on genes that provide a cure, as in the case of a weakened intestinal lining of the host. If we are microbial humans, it makes sense we might enlist microbial doctors.

Microbes on the brink

Are we driving some of our bacteria extinct with our human ways? In the case of pathogens, that might be no bad thing. On the other hand, we have a very powerful immune system; what would it do with itself if there were nobody left to fight? And what about the risk of collateral damage to the good bugs that help us? As a society, we are potentially changing our microbiomes accidently, by altering the environment. As we learn more about our microbiome, we also have increasing options, and motivation, to manipulate it actively. We can try to influence the numbers of good bacteria with our diets and behaviours.

Microbes are so abundant that it might seem odd to talk about being able to drive any one microbe extinct, but in at least a few cases, we might be doing just that. Even the most abundant and widespread microbes might not be safe from our powers to dose the world up on antibiotics. The bacterium *Helicobacter pylori* is named for its corkscrew, or helix, shape. It lives deep in the lining of our stomach in acid as strong as that found in car batteries. This microbe causes the dreaded stomach ulcer, the pain of which can now be remedied, not with milk, but antibiotics.

H. pylori remained undetected until surprisingly recently. It was identified in 1982. Barry Marshall and Robin Warren found it in patients with chronic gastritis and gastric ulcers, conditions previously unlinked with a microbial cause. At the time no one believed a living organism could survive the corrosive nature of the stomach. The discovery of a novel species of microbe lurking right there rocked the medical world. For their discovery of its causal link with disease, the pair received the Nobel Prize in Physiology or Medicine in 2005.

Following this, the bug was linked to stomach cancers. Despite its dark side, *H. pylori* is championed as a hero not a villain by Martin Blaser, a physician and microbiologist at New York University. Blaser has been studying this bug since the mid-1980s. He points out that 80 per cent of individuals infected with the bacterium are asymptomatic and postulates that it may play an important role in the natural stomach ecology. He believes this, and other lines of evidence, show we depend on microbes like *H. pylori* to regulate various functions, including metabolism and immunity.

Blaser says the Western microbiome is already impoverished.[148] Sanitation, demographics, and antibiotic usage are causing a huge decline in *H. pylori,* which was once present in every adult human. He is convinced it is an endangered species. Only 50 per cent of the world's population still harbour *H. pylori* in their upper gastrointestinal tract. In the US, fewer than 10 per cent of children have the bacterium, but when it is present it is still the single dominant species. Perhaps we can live without this ancient denizen of our stomach? We might soon have to. We are changing our microbial ecosystems in ways we do not yet understand and that might be irreversible.

Genomic donations

The microbiome can get sick. Organs are discrete units of our bodies that serve a specialized physiological function. By this definition, the trillions of microbes in our guts are an organ, one we acquire rather than one we are born with. Like any organ, our microbiome can succumb to disease. One time we see our guts going into 'organ failure' is in the case of ulcerative colitis. Patients with this disease suffer terrible cramps and get bloody runs due to colons riddled with painful, bleeding ulcers. There is, however, a way to treat this terrible disease; the 'drug' is human faeces.

Faecal transplants hold great promise but they are not novel. Thousands of years ago, the Chinese had a cure for intestinal ailments, called the Yellow Cure, that involved giving patients faecal matter to drink. Horse handlers also have similar cures. Observations of the natural world reveal many cases where the young eat the faeces of adults, as in the case of elephants.

Faecal transplants are gaining credibility. Marie Myung-Ok Lee recounted her story of 'Why I donated my stool' to the *New York*

Times in the hopes of inspiring others.[149] She was humbled by her own microbial health in the face of the grave illness of her friend. She had great pride in helping provide the cure. The therapy was dealt with at home, DIY-style, under advice from one of the leading doctors in the country. Fresh faecal matter was collected and injected directly as an enema. In the case of Lee's friend, the effect was surprisingly immediate, but wore off. After a few hours of relief, painful cramps would return. Only after a substantial course of transplants did her healthy microbiome take root in him—his ulcers disappeared and health was restored.

To the consternation of some, the US Food and Drug Administration (FDA) moved to regulate human faeces as a drug in May 2013, but subsequently suspended the regulation pending review. Certainly, there are ways to perform the transplants that increase the likelihood of success and the FDA has a duty to ensure safety. It turns out that there is a lot of variation in faecal samples and how they are prepared. This can influence the efficacy of the treatment. In February 2014 Eric Alm of MIT and colleagues argued that a better approach is to regulate faeces as a tissue like blood.[150]

However it is eventually regulated, the logic of this treatment makes perfect sense. Our guts are a living ecosystem. They need the right combinations of species to work properly. Alm and others are working towards understanding the microbial active ingredients so that communities can be tailored specifically in second generation microbiome therapeutics. In the meantime this treatment—relatively easy to perform at home following instructions freely available on YouTube—will certainly spread and could save many bouts of misery in the future.

If you are a healthy eater, and regular, you could be a prime candidate for microbiome donations. Just as when picking a sperm donor, recipients are looking for the best genes.

2 per cent of pandas

All animal species on Earth have microbiomes. Plants do too, and the special coating of microbes that covers plant roots is called the rhizosphere. We are just starting to learn about the microbes that make visible life possible. One fascinating example of how important microbiomes can be to the evolutionary fate of a species, and even to us as humans, comes from genomic studies of conservation icon *Ailuropoda melanoleura*—the giant panda.

Only 2,500 of these shy and solitary animals inhabit the remote regions of China's mountainous bamboo forests. Human population pressures, the destruction of habitat, low fecundity rates, and a restricted diet of increasingly sparse bamboo are all driving this great creature to extinction.

A national symbol of China, the panda genome was sequenced as a flagship project of the BGI.[151] At 2.4 billion base pairs long, it is somewhat smaller than the 3 billion human genome. Unlike humans, however, panda genes were almost unknown prior to the panda genome project. Previously, there were only 27 panda genes in GenBank, one of them the SRY (sex-determining region located on the Y chromosome). Ironically, it was one gene not found in the newly deposited set of more than 19,000 panda genes; the genome sequence was from a 3-year-old, captive-born female.

Why would a bear with sharp teeth and claws choose to feast on a plant that is almost impossible to eat? The researchers examined the gene repertoire for clues as to why the panda looks like a carnivore yet plays the role of humble vegetarian. They discovered a mutation that likely turned off a gene responsible for 'umami taste'—a key part of the enjoyment of meats, cheese, broth, stock, and other protein-heavy foods. They also catalogued genes involved in digestion, finding them in pandas by comparing them to known enzymes in dogs. It turns out that pandas still have enzymes for digesting meat even though they gave the habit up a long time ago. Not so surprising, given that pandas descended from carnivores and will occasionally still hunt. More perplexing was the absence of any digestive enzymes required to eat their only food source—bamboo.

Clearly pandas survive, so something must be breaking down the bamboo's cellulose in their guts. If the enzymes are not produced by panda cells then perhaps they call on gut bacteria to do the job. A follow-on study of 17 wild and captive pandas confirmed that while panda microbiome diversity was unexpectedly low, it contained ample bacteria possessing cellulose-digesting pathways.[152]

A full-grown panda must eat 12–38 kilograms of bamboo a day, or about 10 per cent of its weight, to survive. This is not an easy task. The panda will patiently pull down poles, strip them of leaves with the help of their special 'thumbs', chew them with unusually strong jaws, and pass the fibrous cud past a reinforced oesophagus to a turbo-charged stomach. The panda does a tremendous amount of work for its meal, but it relies on its gut bacteria to finish the job. The microbiome has made the modern panda what it is: a vegetarian, reclusive bear of the bamboo forest.

BGI has now sequenced the genomes of 2 per cent of all extant wild pandas. While this is only 34 animals, it was the highest coverage of any species in a population genomic study.[153] The genomic data confirmed that pandas arose in the Miocene, 5–20 million years ago. It also revealed evidence of a major population expansion coinciding with a dietary switch to bamboo 3 million years ago. It is likely that this switch was enabled by changes in the panda microbiome. Through their microbes pandas were able to colonize a new habitat. The panda microbiome not only changed the ecological and evolutionary trajectory of this species, it could also benefit mankind in our bid for alternative energy sources.

Planting crops for biofuels is now contentious because it may actually need more carbon than it yields in fuel, but using plant waste, such as corncobs or husks, still offers a renewable source of biofuel feedstock.[154] The hard parts of plants are notoriously difficult to break down cheaply and efficiently. The panda microbiome may hold the secret of energy alchemy; specifically, enzymes that could make new biofuels. It is full of enzymes that can turn plant fibres, like insoluble cellulose and lignocellulose, into sugars and ethanol. Pandas have short digestive tracts for their size, suggesting their microbes are particularly potent. Researchers are isolating bamboo-digesting bacteria of pandas, hoping they might yield up enzymes to create precious fuels.

Pandas have long been a symbol of the beauty of wildlife and the need for active conservation efforts. Now they symbolize the potential of genomics. In the future, they might build us amazing bioreactors. This species has a reference genome, a population genomic data set that includes 2 per cent of the total population, and a characterized microbiome—one being mined for enzymes good at producing biofuel. Even the genome of the panda's food, bamboo, has been sequenced.[155] Sad then, that our best-studied 'natural genomic model species' is now one of so many spiralling towards extinction.

The last prairie

The Earth has a microbiome. It exists in the soils, waters, and air of the planet. It, too, despite its tremendous abundance and ubiquity, is rapidly changing. As with global climate change, there is evidence of a human role in the changing Earth microbiome. Take soils for example. The productivity of soils is thanks to their indigenous microbes. We change these communities when we start to manage our soils, like in our gardens and of course on our farms, especially industrial-scale ones. The relatively gentle tilling of soil might have a similar impact on soil microbes as clear-cutting a forest does on larger species.

When we raze natural land to plant commercial crops, through tilling we remove the surface layer of life, but we also significantly alter the vast ecosystem of invertebrates and microbes within the soil. With the loss of the indigenous plant communities, so too go associated animals, like pollinating insects, and the subterranean life. Soils have a microbiome that performs a range of biological functions including helping plants to grow. Just as our gut microbiome changes in response to what we eat, the soil microbiome is profoundly changed by modern farming practices.

The Great Plains of North America is one of the most fertile places on Earth. The 'Wild West' is epitomized by visions of giant herds of peaceful bison roaming the vast open grasslands. With a growing human population, these herds disappeared—hunted for food, clothing, and shelter. Whereas bamboo remains even as the pandas disappear, the tall grasses that formed the foundation species of the prairie have largely gone—replaced with another grass preferred by humans, wheat.

The prairie once covered 10 per cent of the US, stretching over more than 150 million acres. It covered a strip down the middle of the US from Minnesota to Texas. Progressively tilled after the European settlement of the Midwest in the mid-19th century, it is still extremely productive thanks to continued tilling and intensive fertilization.

Beyond the loss of the tall grass, tilling took out hundreds of species of flowers, the surface creatures that fed on them, and untold masses of underground life. Noah Fierer, at the University of Colorado, wanted to explore how tilling impacted prairie soil microbiomes.[156] He collected tilled and untilled soil samples from 31 pairs of sites across the prairie's historical range. The task proved harder than it sounds. Despite the vast range of original prairie, Fierer's team found it very difficult to find sites that had never been tilled. Many of the samples came from cemeteries.

Modern terraforming of the prairie soils is evident. Fierer's analysis revealed that communities varied between geographic regions and diversity was driven by environmental and climatic conditions, such as rainfall. They also differed in the abundance of a group of bacteria known as Verrucomicrobia. Fierer is a champion of these mysterious bugs, once thought to be quite rare for artefactual reasons but now shown to be less so after Fierer developed better DNA technologies for detecting them.[157] He and his colleagues studied 181 diverse soil samples from Antarctica, Europe, and the Americas with improved PCR primers and showed that Verrucomicrobia could be detected in 180 out of 181 soils examined. Estimates of Verrucomicrobia jumped from 7 per cent to 23 per cent of soil microbes on average. The relative abundance of Verrucomicrobia was highest in grasslands reaching >50 per cent. So, why are they there?

Fierer believes that Verrucomicrobia are critical to the functioning of tall grass prairie soils. He is fascinated by the fact that this one group dominates these productive soils and yet we don't know why it is so abundant or exactly what it is doing. Only a few species of these bacteria have been described to date and they remain poorly understood because they are so difficult to culture in the laboratory. One clue might be found in their capacity to break down complex carbon-containing structures, such as the roots of prairie plants. They also seem skilled at living in nutrient-poorer soils and are likely out-competed by other species in fertilizer-rich soils.

Fierer and his team used the data from the 31 precious sampling sites to extrapolate expected historical microbial communities across the full prairie. This microbial map will hopefully aid in restoration projects and will inspire similar studies of other ecosystems. If human faecal transplants can help restore healthy microbiomes, perhaps soil microbial transplants could do the same.

Certainly, the Great Prairie study offers more evidence of how substantially we are changing the planetary genome, at the very point we are learning to map it. It shows our power to terraform to destroy, but perhaps also hints at our future abilities to genomically terraform to restore biological balance. It also highlights the need to understand genomes in context right up to the ecosystem level.

6

TERRA-GENOMING

The lingering kiss

A recent survey found that as many as 53 per cent of Americans and 59 per cent of Europeans agreed with the patently wrong statement: 'Ordinary tomatoes do not contain genes, while genetically modified tomatoes do.'[158] In an interview about his book *Life at the Speed of Light*, Craig Venter lamented that it was hard to communicate the promise of genomics when half of the public don't seem to know that a tomato has DNA.[159] If these rather shocking statistics reflect a general truth, the mainstreaming of genomics into modern culture will hopefully soon remedy such ignorance.

All cells have genomes, with few exceptions. Red blood cells jettison their nuclei—the sac that holds DNA—when they mature. This enables them to adopt the concave shape they need to effectively hold oxygen. On the other hand, reproductive cells, eggs and sperm, carry only half a full payload, just a single 'haploid' set of 23 chromosomes. The sperm's sole mission in life is to deliver this genetic cargo to the egg and restore a full genomic complement.

Since all cells have DNA, DNA bathes our surroundings. We eat genes every day. Food made from any meat, vegetable, or fruit, drinks like coffee and tea, or spices, like cinnamon, nutmeg, and pepper—all have DNA. Wine and beer contain DNA from the yeasts used to ferment alcohol as well as from the grapes or hops. One can leave DNA for up to an hour in a lover's mouth after just a bit of passionate kissing.[160]

There is DNA all over our houses and workplaces too, in the leather of our clothes, the paper in our printers, and the wood in our furniture. If that were not enough, we slough off cells all the time, leaving a trail of DNA behind us. House dust is largely composed of dead skin cells, and as a consequence, our homes are covered in our DNA, as well as the DNA of our cats, dogs, and any other furry friends we choose to live with.

In 2011, scientists found that a significantly high proportion of supposedly non-primate genomic databases were in fact filled with a lot of primate sequences. A search for primate-specific repetitive sequences, called Alu sequences, found contamination was rife—most likely human DNA from those who prepared the biological samples for sequencing. Perhaps it is no surprise to find our sticky fingers in the genomic cookie jar, given the ease at which DNA is spread and its persistence. DNA is hard to wash off. Security firms are now using it in paint packs that burst when touched, to cover thieves, not in indelible dye but invisible, long-lasting 'DNA tags'.[161]

The ubiquity and persistence of DNA is a boon for field biologists. The risks of contamination notwithstanding, the stability of DNA and the potential to recover it from tiny samples is revolutionizing field biology. It is particularly useful for those studying difficult-to-capture, rare, and endangered species. A sample of blood or a snip of tissue can yield plenty of DNA for study with no lasting harm to the organism. For example, whales are not the easiest organisms to coax into the lab, and butchering them in the name of science is repellent to most researchers. Fortunately, scientists now have a powerful alternative: simply fire a dart from a boat to sample a small skin plug that will yield sufficient DNA for many analyses without harming the animal.

Other methods require no direct interaction with the organism at all; one can simply collect the DNA that animals shed in hair follicles, feathers, horns, faeces, or other bodily parts or excretions. Even the foot mucus of snails can yield DNA when they crawl over a special DNA collection card. Chemicals in the card lyse the cells in the mucus, denature the proteins to immobilize them, and fix the DNA for preservation. Once dried overnight, the DNA can be kept at room temperature for months or years. Even the most scattered, diffuse, or diluted genetic leftovers can remain detectable and scientifically useful for considerable periods of time.

Probably the most significant source of DNA around us is the invisible world of microbes. As we are learning from the explosion in microbiome studies, bacteria are in us as well as all around us. They inhabit our most intimate niches, and as they evolve so rapidly, they quickly develop unique signatures associated with their host or micro-habitat. Molecular traces left on keyboards, for example, can now identify past users through sequencing the skin microbiome they leave behind.[162]

DNA is everywhere and our biological universe runs on DNA. Humans are part of this dynamic, affected by the movement of genes through space and time. Increasingly, we aren't just consumers of nature's software. We have become adept software developers in our own right. We shuffle the planetary genome around to suit our needs for food, shelter, security, and energy. We have been doing this since the dawn of mankind, through hunting and gathering, but the process accelerated with the agricultural revolution, gained pace through the industrial revolution, and recently went to warp speed thanks to globalization and computers.

Now we are terraforming wholesale. We are terra-genoming. We are changing the distribution of DNA on the Earth as never before. Prior to learning how to read and write DNA, we just manipulated the organisms that contained it. As we explored the world, we picked up a few useful species, and many more stowaways. In addition to the domesticated species, we unwittingly assembled an ark of travelling companions, from rats and mosquitoes to microbial pathogens. We were a veritable travelling circus, accompanied by a swarm of biodiversity, some essential to our peregrinations, others the source of untold misery. As we travelled and settled the world, we manipulated the genetic code underpinning Earth's life support systems.

We can help reconstruct the history of this transformation with the aid of DNA itself. The universality and stability of DNA, coupled with affordable technologies to read and write this code, help us to understand the natural world. Organisms, dead or alive, are delivering up their DNA, from cheek swabs to faeces to ancient bones. We are using the data to decipher the history and movements of life on Earth.

Reunion

In an epic story of bravery and ingenuity, humans have colonized every continent. Nobody documented the earliest human odyssey, but we can read elements of it in our DNA, as well as the DNA of the species that travelled with us. The last chapter of this incredible journey occurred in the Pacific. Arguably, the most stunning feat of Earthly exploration, it is eclipsed only by our modern voyages to outer space.

Humanity streamed out of Africa and colonized every corner of the planet. Leaving Africa about 60,000 years ago, modern humans fanned out across Asia. One of these groups eventually crossed a 1,000-mile-wide land bridge from Siberia to Alaska, some 15,000 to 20,000 years ago, and thence all the way to Patagonia in South America. Rising sea levels subsequently flooded the land bridge to form what we now know as the Bering Straits. The Americas thus became one huge island with the Pacific on one side and the Atlantic on the other.

Humans had become adept at travelling over land and along waterways, even making some short sea crossings, but ocean-going vessels and long-distance navigation awaited the evolution of new technologies and know-how. Consequently, islands in the Pacific, Atlantic, and Indian Oceans remained unpopulated for most of human history—until Polynesians conquered the Pacific and Europeans became global explorers.

The Pacific is the world's largest geographical feature, covering about a third of the globe. If you centre your view of Google Earth on Tahiti, you will get a feeling for its magnitude—almost no land is visible save a few large islands and slithers of California, Australia, and Antarctica. Zoom in, however, and you will see that the endless blue is a night sky, a constellation of many (poly) islands (-nesia).

Where did Polynesians come from? One line of evidence in particular, the sweet potato, suggested that the islands were populated from South America. Sweet potatoes are indigenous to the continent of South America but are now cultivated across the islands of the Pacific as a staple food. The sweet potato was domesticated in the Peruvian highlands some 8,000 years ago, so scholars guessed that Spanish and Portuguese explorers introduced the crop to the Pacific Islands in the 16th century. Yet, data suggested that this might not be true. For example, the oldest carbonized sample of the crop found by archaeologists in the Pacific dates to about 1000 CE—nearly 500 years before Columbus' first voyage. What's more, the word for 'sweet potato' in many Polynesian languages closely resembles the Quechua word for the plant.

So might prehistoric South Americans have introduced the sweet potato? Are Polynesians descended, at least in part, from Americans? This was what Thor Heyerdahl believed, and he constructed the most famous raft in history, the Kon-Tiki, to demonstrate it was possible. We now know that in fact the truth was even more amazing: Polynesians voyaged to South America—and back.

Polynesia's epic tale has been reconstructed using several lines of evidence, most recently including DNA. In particular, molecular anthropologists have learned a lot from what they call the 'canoe biota'—the organisms carried deliberately or accidentally in the twin-hulled Polynesian sailing canoes. Genetic analyses of various plants, rats, pigs, and chickens[163] provide some of the best proof thus far that not only did the Polynesians sail out of Asia to colonize the remotest Pacific Islands, but that they even made it to the shores of South America.

A study published in 2013 by a group of experts in human genetics, migration, and the analysis of ancient DNA, for example, uncovered further evidence.[164] The team studied the now extinct Botocudo Indians of Brazil. Using DNA extracted meticulously from ancient skulls, they report the identification of mitochondrial sequences belonging to haplogroups characteristic of Polynesians.

The sweet potato mystery was finally unravelled using DNA by a team of scientists at France's Centre of Evolutionary and Functional Ecology, and at CIRAD, the French agricultural research and development centre.[165] Their analyses required samples of the plant from before modern trade had obscured the trail, mixing up the prehistoric patterns. Luckily, such samples were at hand thanks to Joseph Banks, the gentleman botanist who accompanied Captain Cook on his first voyage to the Pacific; two of his sweet potato plants were found in the herbarium at Kew Gardens.

It appears that the sweet potato was transported from its native range at least three times: first by Polynesians voyaging with it back from South America, then by Spanish traders sailing west from Mexico, and finally by Portuguese traders coming from the Caribbean. The Spanish and Portuguese varieties ended up in the western Pacific, while the older South American variety dominated eastern Polynesia, explaining the observed genetic differences. DNA analysis, not of humans but of one of our staple foods, helped reveal one of the greatest feats of human bravery.

Unicorns

DNA is helping us to recount the prehistory of human migration. It also enables us to catalogue and track all the other species on Earth. Every organism comes with a full DNA name written inside; that is, its genome. Calibrating our genomic tools correctly, we can set our analytical sequencing power to the level of 'species' and read out just enough of the name to tell if it is 'giraffe', 'elephant', or 'horse'. Reading these indelible names, DNA detectives can determine the species—human, hippo, or hydrangea—of any organism, or part thereof.

A global species DNA naming convention, one that Linnaeus would have deeply welcomed, was masterminded by Paul Hebert of the University of Guelph. In 2003, the Canadian zoologist proposed a massive effort to build a global database of species names linked to genetic sequences of standardized regions of DNA, known as DNA barcodes.[166] The idea went viral and the International Barcode of Life (iBOL) initiative now claims to be the largest biodiversity genomics initiative ever undertaken.[167] As of early 2014, iBOL has amassed coverage for almost 3 million specimens and 200,000 named species in its Barcode of Life Database (BOLD).

Examples of how this 'yellow pages' of life is helping us to understand biodiversity abound. In the deepest forests of Vietnam DNA-barcoding of leeches is being used to track an elusive creature called the Asian unicorn. A type of ox (but having two horns), it was first photographed in 2013. Thomas Gilbert of the University of Copenhagen and colleagues looked to leeches to help track this elusive animal.[168]

Gilbert's team tested the blood meals of 25 wild leeches and found DNA in 21. Moreover, they showed that prey DNA can be preserved for up to four months in a leech stomach. Leeches know where to find the unicorns when humans can't. They, like other micro-predators, including ticks and mosquitoes, are now being used to track rare and endangered species around the world.

The living or the dead can be tracked. Demian Chapman, a biologist with the Institute for Conservation Science at Stony Brook University, and collaborators collected samples of shark fin soup from 14 American cities and used DNA analysis to study the species in shark fin soup.[169] Eight species of sharks, including blue, shortfin mako, bull sharks, and scalloped hammerhead, were identified, showing it is possible to get DNA from highly processed foods like soup. It was all the more surprising given that shark fins are dried in the sun, shipped long distances, and chemically treated.

DNA barcoding is even weighing in on religious law—by determining whether unwanted worms broke kosher law or not. Kashrut is the body of Jewish law dealing with what foods are kosher and how they must be prepared. The codex states that 'of the things that are in the waters', anything with fins and scales may be consumed so long as the flesh has not been mixed with intestinal remains. When rabbinical experts of the Orthodox Union noticed worms in their sardines they feared they might be from intestines. DNA experts at the American Museum of Natural History in New York determined that it was safe—they identified the nematode worms as anisakine species that 'do not inhabit the intestinal lumen'.[170]

Even complex mixtures of species can be disentangled. Eske Willerslev, an expert on ancient DNA from the University of Copenhagen, showed that next-generation sequencing technologies are capable of reading short pieces of degraded DNA. Using this high-throughput technology his team obtained DNA sequences from a range of previously 'DNA-less' places. Overnight, even ancient samples, for example of soil, became 'readable'. Willerslev had invented what he calls 'dirt genomics'.

Willerslev's team pioneered the concept by reconstructing a Pleistocene community from soil.[171] Five permafrost cores ranging from 400,000 to 10,000 years old were shown to contain 19 different species, including mammoth, bison, and horse. It was now possible to study the life of regions that lacked macrofossils—by looking for DNA.

Today, Willerslev is setting 'DNA traps'.[172] He has placed them in well-characterized safari parks and farms to test their accuracy: are they able to reconstruct a species list that corresponds to what is actually in the park or farm? They extracted DNA from the test site's soil, sequenced it, and compared it to sequences in the global reference database GenBank. The results were more than encouraging. They found all the animals they expected—with the exception of an animal that was recently introduced to one of the game parks—a type of giraffe. Furthermore, the quantity of DNA recovered for each species roughly reflected the number of animals of that species present, adjusting for body weight.

This technique is now known as eDNA (environmental DNA) and it has great potential for genetic monitoring programmes on land or at sea. As Ryan Kelly and colleagues at the Center for Ocean Solutions in California put it: 'the potential to distil policy-relevant ecosystem-level information from a glass of seawater is new'.[173]

There is an expanding list of practical applications for DNA barcoding and eDNA technology. DNA barcoding is widely used in forensics, for example, to test for endangered species at customs checkpoints. It can also be used in field studies in any number of ways including to identify samples that cannot otherwise be easily identified by taxonomists, like plant leaves even when flowers or fruit are not available, or insect larvae which often lack the diagnostic characters of adults, or to determine the diet of an animal based on a DNA analysis of stomach contents or droppings. It can be used to identify any number of products in commerce, such as the exact species of plants in tea bags, or whether or not there is horsemeat in your meatballs. All this is very useful information for assessing food quality and legality. Increasingly, we can track any living thing, both the quick and the dead, by DNA.

Invaders

Humans are changing the distributions and ranges of species, often to our detriment. Invasive species are a global problem. Such species end up in the wrong place at the wrong time. They become devastating pests and weeds when introduced to novel ecosystems. These 'alien invasive species' cause massive disruptions that can lead to long-term changes, including the extinction of other species. Invasive species are like criminals on a 'most wanted' list—in fact there is a top 100 invasive species list[174] to aid in their management. Great effort is going into learning the 'DNA names' of all species that are known invaders.

Scientists are watching the 7 billion dollar fishing industry of the Great Lakes with great concern and have a specific invader in sight—Asian carp.[175] Its name strikes fear into the hearts of environmental agencies and fisherman alike. Worst case scenarios suggest that the fish, swimming up from established breeding populations in the Mississippi, could threaten boating and fishing industries, undermining the livelihoods of thousands by destabilizing the Great Lakes ecosystem.

As a defence, an electric barrier has been erected in a shipping canal 37 miles from Chicago to block their path towards Lake Michigan. Just one live Asian carp has been found beyond that point, although numerous DNA samples have turned up past the barrier and in Lake Erie. The first positive DNA hit came from Lake Michigan's Calumet Harbor in 2010; a second was reported in 2013.

Presence of DNA in the water is not evidence that an adult fish was actually living in the lakes, or that a breeding population has or will establish. Carp DNA could have come from bird droppings, as the species is already present in some of the region's rivers. Nevertheless, tracking eDNA is like using a smoke detector: it raises the alarm but one still needs to find the fire, ascertain the level of danger, then decide what to do about it. Genetic monitoring offers an early-warning system.

Insects are some of the most problematic invasive species. They can threaten public health. Humans have spread the mosquito *Aedes aegypti*, for example, around the world and along with it the risk of dengue fever has risen substantially. Countries that were previously free of the disease now have an additional burden on their health services. Novel approaches to mosquito control are leveraging genomic technologies. These 'genetic control' programmes can involve modifying either the mosquito genome or its microbiome. In some cases, they aim to suppress the mosquito population enough to break the disease transmission cycle. In other cases, they can drive a new genome through the natural mosquito population that renders them incapable of carrying the disease organisms. That leaves the mosquitoes as a biting nuisance, but at least they will no longer make you sick.[176]

Invasive insects are also a menace of agriculture. The Mediterranean fruit fly (medfly), for example, is one of the most feared potential pests in California. Enormous effort is expended to try to keep this species out of the state. One approach is to drop 125,000 sterilized male medflies per square mile on regular bombing missions flown out of airports in Southern California.[177] The hapless males are grown in massive fruit fly factories in Guatemala and Hawaii. The idea is that if any female medflies do make it to California, chances are that they will mate with a sterilized male and so produce no offspring.

Raising large number of medflies is a challenge. Especially because, while they are needed to breed, one does not want to release females into the wild as they attack fruit to lay their eggs—the reason fruit farmers don't want the medfly in the first place. Again, genomic technology is offering novel solutions. One technique involves inserting a gene in the factory strain that kills female flies if the temperature in the factory is raised.[178] With the females thus removed, the male survivors are sterilized, often using radioactivity, and shipped to 'bomber command'.

Monitoring the effectiveness of this programme is a challenge. If you find a female, how can you tell if it was mated by a sterilized factory male or by a wild type invading male? Another ingenious transgenic solution has been developed. It inserts a fluorescent jellyfish gene into the factory-bred medflies. This DNA not only gives the factory males a new 'DNA nickname', but it is easily detectable in a mated female, as their sperm will glow in the dark under ultraviolet light.[179]

Such advances are not going to solve all the problems that will come from displaced biodiversity but they will help avert the worst consequences of globalizing the planetary genome.

Genes on the move

Microbes notoriously pick up and integrate DNA into their genomes. This is how antibiotic resistance races across the world. The ability to pick up DNA from the environment and from other bacteria is a key evolutionary adaptation in bacteria. It's a variation of sex that allows at least some recombination of genetic material between the genomes of clonal organisms. This 'transformer' ability of bacteria is widely recognized, and not surprisingly it happens in our guts.

'You are what you eat', the cliché goes. In 2010, Jan-Hendrick Hehemann led a study that showed that not only do microbes break down DNA to its component molecules during the digestion of food in our guts, but they also have another trick: they can grab a stretch of DNA from a food item and add it to their own genome.[180] Justin Sonnenburg called it 'genetic pot luck':[181] 'our microbes are bathed in a constantly changing soup of foreign genes within our digestive systems'. Grabbing genes is like buying a lottery ticket and hoping for a prize.

Our gut bacteria help us digest our food. When we lack the enzymes for breaking down a particular kind of molecule it just passes through us. You can drink flakes of gold in Goldschläger, and they pass right through. Among real food items, there are a surprisingly large number of molecules we can't digest in plants. Our bacteria, though, have the right tools (genes) for the job. Bacteria in our guts specialize in producing 'carbohydrate active enzymes', called CAZymes.

These genes come in many sizes and shapes but all produce enzymes that chop up the complex sugars found in plants, making their energy accessible to the human digestive system. These genes are absent in the human genome and only found in our bacterial friends. *Bacteroides thetaiotaomicron* has an amazing repertoire of 261 CAZyme genes and is extremely adept at transforming our food into energy for us. Unfortunately, it is sometimes too good—it is particularly prevalent in obese people.

Hehemann's study included a bacterium called *Zobellia galactanivorans*. It had a CAZyme that chops up porphyran, an abundant sugar, or polysaccharide, found not in land plants but in seaweeds of the oceans. Hehemann pondered why it would appear in the human gut but a search revealed it wasn't just a fluke of his sample. Hehemann found it, as expected, in marine bacteria, but also in other human microbiomes, specifically from individuals in Japan.

Could it be a gene selected to help digest seaweed, a staple food in Japanese cuisine? The authors surmised that seaweed, prevalent in the Japanese diet as nori and used to wrap sushi, was probably the source of the microorganisms that introduced the genes. They suggested that non-sterile food, like raw seaweed, could carry natural environmental microbes and be a factor in generating CAZyme diversity in human gut microbes. In this scenario the algae arrived covered in the enzymatic machinery needed for its digestion and were adopted by the resident microbiome.

We are also mediating the exchange of deadly genes that threaten our lives. One of the biggest health threats facing the planet today is antibiotic resistance. Antibiotics are one of the primary reasons why more of us live longer. The power of microbes to evolve adaptive strategies that neutralize and evade antibiotics has led the World Health Organization (WHO) to declare antibiotic resistance one of the biggest health threats to humans today. Much has been written about the threats of antibiotic resistance and 'superbugs'.

We provoke these otherwise benign microbes at our own peril. Microbes have been battling nasty chemicals in the environment long before the advent of human use of medicinal antibiotics and are highly adept. Antibiotic resistance genes evolve and spread as bacterial counter-defences. They occur naturally and derive from new mutations in ancient genes that bacteria possess because they confer resistance to dangerous chemicals. Fungi naturally produce them to kill off bacteria.

Like yin and yang, exposure to antibiotics drives the faster evolution of resistance, either through novel mutations or more frequently the spread of existing genes. The modern world is flooded in antibiotics, from sources as varied as soap to the feed we give livestock. A surprising number of antibiotics, like other pharmaceuticals, end up in the environment.

Many drugs are excreted from our bodies without being metabolized and flushed down the toilet. Interestingly, studies of cocaine in the River Thames show a peak of concentration around Westminster in London—home to the British parliament.[182] Environmental scientists worry about the impacts of these molecules. Junkie frogs are not a top concern, but feminization of male fish by oestrogens excreted in birth control pills is worrying. Entire fish populations are endangered because exposure effectively neuters the males.

Active compounds are also released as by-products of the manufacturing process. A majority of the world's pharmaceuticals are manufactured in China and India. The Patancheru industrial area is a hotspot of biomedical research and factory-scale pharmaceutical production. Nearly 90 drug manufacturers are believed to pass effluents into one incredibly polluted river.

Initial studies led by Joakim Larsson in Sweden showed that the river contained drugs at therapeutic levels; that is, at concentrations you would find in the bloodstream of a patient. This drove his curiosity about the metabolism of the microbial communities that might be living in these waters. His subsequent work revealed that bacterial diversity dipped fractionally downstream from the treatment plant.[183]

More interestingly, he confirmed his suspicions that these bugs harboured huge numbers of antibiotic resistance genes. Human carelessness had created a genetic time bomb. An immense 'resistome' lurked in the drug-infested waters. Luckily, many of the bacteria present in these rivers are only found in the environment and do not infect humans. Still, researchers speculate on the dangers of their genes passing into pathogens that do infect humans, especially as these waters are used for irrigating crops.

It is fortunate that 'wild type' bacteria often outgrow antibiotic-resistant strains. Without persistent selection the resistant genes tend to disappear from bacterial populations. Still, this study offers dire proof that antibiotic resistance genes are circling in the Earth system—especially terrifying since novel antibiotics have not been found for decades.[184]

A dead sea comes to life

It was in 1676 that Antonie Philips van Leeuwenhoek, a Dutch inventor, innovated a way to grind exquisitely fine glass lenses. With them he created the first microscope and fell into the invisible world. He was astonished at the ubiquity of single-celled creatures, which he called 'animalcules'. He peered at them in everything from the scrapings off his teeth to rainwater. His curiosity was insatiable and he was rewarded with a stream of discoveries, including the Protists in 1674, which he called 'Infusoria', the Bacteria in 1676, and spermatozoa in 1677. It was a golden age of exploration; everywhere he looked with his new lens, miniature miracles of biology came into view.

The advent of DNA technologies proved a boon for microbe discovery. The pioneering work of Carl Woese led to the discovery of a Third Domain of Life, the Archaea. Woese used DNA analysis of the 16S ribosomal genes.[185] He continued to hunt down novel microbes using 16S, which became the commonly used 'DNA Name' for Bacteria and Archaea. His colleagues, most notably Norman Pace, innovated methods of pulling 16S ribosomal genes from environmental samples—homogenates of soil, water, or the tissue of larger organisms.

This started the cataloguing of microbial life in earnest, charged by an interest in genetic 'bioprospecting'. Interest in extremophiles, species living in unusual environments like hot springs, was particularly intense because of hopes of finding commercially important enzymes to replace more expensive or environmentally damaging non-biological technologies. Ironically, the most famous success story is one that underpins the modern DNA revolution itself. The polymerase, or DNA copying enzyme used in the Nobel Prize winning invention PCR, comes from a thermophile living in a geyser in Yellowstone Park, *Thermus aquaticus*.

Jo Handelsman coined the term 'metagenomics' in 1998[186] to mean chopping up the DNA from environmental sources and placing it into bacteria where it could be grown up in the laboratory. This was a revolutionary new approach designed to determine the functions of microbes. New genes could be tested en masse in the lab for the presence of interesting traits like antibiotic resistance or the ability to break down a particular type of molecule. In 2005, Craig Venter launched his Global Ocean Survey with a study of water from the Sargasso Sea and changed both how metagenomics was done and our view of the oceans and microbial diversity for ever.[187] Due to its success, metagenomics would thereafter largely shift to 'shotgun sequencing' of whole communities.

The Sargasso Sea is comparable in size to the United States, and is named for the ubiquitous seaweed, *Sargassum*, that floats lazily atop its surface. At its heart lies the Bermuda Triangle, the legendary place with the power to mysteriously engulf ships and aeroplanes. The Sargasso Sea is a 'sea without shores' floating within the Atlantic. Its waters are isolated from the surrounding ocean by currents that run around its perimeter; without access to mineral and nutrient-rich runoffs from terrestrial coastlines, the Sargasso Sea is nutrient-poor, or oligotrophic. Without nutrients, there is less planktonic life, which gives these waters their distinctive deep blue colour and exceptional clarity. Underwater visibility is up to 200 feet but the bottom lies miles below the surface on the Nares Abyssal Plain.

The Sargasso Sea seemed a perfect place to trial the new shotgun sequencing of natural environments. As diversity in the barren waters of the Sargasso Sea was expected to be low, it should have been more tractable. But life was everywhere. A veritable treasure trove of new evolutionary lineages and never-before-seen genes emerged. Venter and his team discovered 1,800 species of microbes, including 150 new ones, and over 1.2 million new genes.

In the best tradition of maritime exploration, Venter's yacht, the *Sorcerer II*, circumnavigated the globe stopping every 200 kilometres to collect water samples. Analysis of the first 41 stops produced

17 million new genes and doubled the size of the world's public DNA databases. The Global Ocean Survey exceeded all expectations for detecting novel microbial diversity.[188]

The race to sequence the Earth, and its oceans, was on. Researchers began the mapping of habitats around the globe to ask 'who is there' and 'how they contribute to the functioning of natural systems'. Just like the microbiota of our guts, microbes do important jobs 'for the planet'. Not least, photosynthetic microbes seem to account for about half of all the energy that life on Earth derives from the Sun.

In the marine environment, the 'International Census of Marine Microbes', led by Mitch Sogin and Linda Amaral-Zettler of Woods Hole and Jan de Leeuw of the Royal Netherlands Institute for Sea Research, carried out a groundbreaking census of marine microbes that vastly expanded upon what Venter had achieved.[189] Since then the most ambitious study of microbes in the natural environment is targeting the entire planet: the Earth Microbiome Project[190] led by Jack Gilbert, Rob Knight, and Janet Janssen. This project has already catalogued some 9 million different types of bacteria by examining DNA from a huge range of habitats. The Earth is a microbial planet and characterizing and understanding the roles of the invisible majority will be one of the largest contributions of genomics.

Shock and awe

We are blindly transforming a system on which our prosperity depends. These DNA stories show us that we are terra-genoming the Earth at an increasing pace, and we have a limited understanding of the implications of such changes. In fact, we are still stuck at the inventory stage—only just beginning to learn how much life there is on Earth.

So just how many species are there? Ecologist Lord Robert May drew fame for asking this question.[191] To his dismay, he is still awaiting an answer. In a 2010 article he asked: 'If some alien version of the Starship Enterprise visited Earth, what might be the visitors' first question? I think it would be: "How many distinct life forms—species—does your planet have?"'[192] Embarrassingly, he reminds us, our best-guess answer would be in the range of 5 to 10 million eukaryotes, never mind the viruses and bacteria, but we could defend numbers exceeding 100 million, or as low as 3 million.

Most life has not yet been described. In 2011, Camilo Mora of Dalhousie University, Canada and colleagues produced an estimate of global diversity. They argued that the higher-level groupings of taxonomic classification followed 'a consistent and predictable pattern from which the total number of species in a taxonomic group can be estimated'.[193] From this they estimated almost 9 million eukaryotic species on Earth, of which some 2.2 million are marine species. Their numbers broke down as approximately ~7.77 million animals, ~298,000 plants, ~611,000 fungi, and ~36,400 protists.

This narrows the playing field but is still an underestimate. It doesn't include the Bacteria or Archaea, the vast majority of genomic biodiversity. Mora's study found that 86 per cent of the eukaryotic species on land and 91 per cent in the ocean still await description. They estimated that, based on a taxonomic 'business as usual' scenario, describing all these species might cost some US$364 billion, 100 times more than the Human Genome Project, and would require 303,000 taxonomists working for more than a millennium.

New species continue to turn up. Quentin Wheeler's International Institute for the Exploration of Species (IIES), formerly at Arizona State University, selects the top 10 of them as the 'shock and awe' species of the year.[194] Recent winners include a bat the size of a raspberry, a glow-in-the-dark cockroach, a yellow mushroom that looks so similar to the cartoon character SpongeBob SquarePants it is named *Spongiforma squarepantsii,* and a harp-shaped carnivorous sponge (see Figure 5).

The vast numbers of unknown species, or dark taxa, are just the tip of the iceberg. With them come unknown genes. 'Unknown' genes dominate most metagenomic studies of microbes, representing one of the great challenges of genomics.[195] Some fall into known gene

Figure 5. Novel species continue to appear at a fast pace and 86 per cent of eukaryotes await formal description. Discovered in 2012, the carnivorous 'harp' sponge, *Chondrocladia lyra*, is found off the coast of California at depths between 3,300 and 3,500 metres (10,800–11,500 feet). It was only found once underwater robots, pioneered and operated by the Monterey Bay Aquarium Research Institute (MBARI), were developed and deployed to probe the seabed as part of MBARI's ongoing attempt to explore this vast remaining biological frontier. This technology has surfaced many other amazing aspects of deep-sea biodiversity.

families, but a large number are 'orphans'. Orphaned genes belong to small, rare, or poorly sampled gene families. Even the human genome still harbours genes that have no ascribed role. It was not just the small number of genes in the human genome that was shocking, but also the general number of orphaned and uncharacterized genes. Similar results were found in sequencing model organisms like the fruit fly too.[196]

Given the scale of the task, one question is whether we have time to name, much less sequence, even a small proportion of species, before they disappear. It is not even known how many species we currently describe every year; less surprisingly, how many we lose—we don't know the exact extinction rate either. Estimates of species descriptions range from 8,000 to 18,000, while it is thought that eukaryotic species are disappearing at more than 1,000 times the background rate.[197]

With these caveats in mind, Mora and colleagues made projections under optimistic and pessimistic extinction scenarios. With a constant rate of 8,000 species being described each year, we are likely to have lost at least 4 million eukaryotic species, about half of today's total, before we manage to describe all of those that remain. That is the optimistic scenario.

Cataclysmic die-offs happen but rarely. Mass extinctions are defined as times when the Earth loses more than three-quarters of its species—again ignoring the microbes—in a geologically short interval. The 'Big Five' extinctions in the history of life in chronological order were the Ordovician, Devonian, Permian, Triassic, and Cretaceous, which wiped out the dinosaurs 65 million years ago.

In 2011, Anthony Barnosky and colleagues discussed whether the Earth's sixth mass extinction might be imminent.[198] Since current extinction rates are higher than background rates and higher even than those that caused Big Five extinctions in geological time, we might pass the critical 75 per cent threshold in as little as 300 years and enter the Anthropocene mass extinction.

Apocalyptic forces act on the planetary scale during mass extinctions crises. The Cretaceous event that brought an end to the dinosaurs is associated with an asteroid strike that left a massive crater in the Yucatán peninsula in Mexico, though the exact causes of the awesome purge of life are still debated. One thing is clear: *Tyrannosaurus rex* and 76 per cent of animals and plants disappeared, seen no more in the fossil record.

The other major die-offs played out under complex scenarios, synergies of unusual events, perfect storms of terrible circumstances. The Granddaddy of all die-offs, the Permian extinction, occurred 251 million years ago. Global extirpation seems to have involved a combination of Siberian volcanism, global warming, marine death zones devoid of oxygen, elevated hydrogen sulphide and CO_2 concentrations, and ocean acidification. At its close, and again not including the microbes, only 4 per cent of species remained.

Ancient mass extinctions were all triggered by geo-climactic catastrophes: nature. We were not involved, or even evolved. Species loss in 'the age of man', the Anthropocene, however, is caused by us. Some deny human activities cause climate change, but it seems more generally accepted that we cause biodiversity loss. We can see it for ourselves in our local communities as well as on TV as exotic locations are stripped of life. We co-opt resources, fragment habitats, introduce non-native species, spread pathogens, kill species directly for food, security, shelter, or thrill, and then there is pollution up to and including greenhouse gases.

Our destructive power is on the scale of a death star; we are the modern equivalent of that asteroid which saw off *T. rex*. How the Anthropocene will play out is a matter of debate, but it appears clear that human actions are pushing Earth's subsystems to dangerous extremes.[199] In the positive thinking camp, there are those who believe we will prevail,[200] while others believe that we are lost.[201]

Regardless of outcome, we have significantly changed the playing field for genomes and genes across the planet. Human population growth has proceeded at a staggering rate since the scientific revolution of the late 17th century—doubling in the last 50 years. Of all the humans born since our species began, this means 1 out of 12 is alive today. It is in this backdrop that we stand to lose 75 per cent of animal and plant species in as little as 300 years. Future human generations will inherit a planet with a significantly changed and deeply impoverished planetary genome.

7

WE ARE ALL ECOSYSTEMS NOW

Quantified Self

Evolution might have made humans self-obsessed to start with, but genomics is letting us take our navel-gazing to new heights. We are set to understand ourselves as never before—from our molecules up—as systems. Michael Snyder of Stanford University was one of the early pioneers, undergoing what is known as integrated personal 'omics' profiling (iPOP). Snyder not only had his entire genome sequenced but also subjected himself to a barrage of other analyses, collectively known as 'omics',[202] over a 14-month period. The scientific smorgasbord included transcriptomics, proteomics, metabolomics, and autoantibody profiles. Essentially the approach takes a large-scale look at all the molecules that make us tick. The result was what Eric Topol of the Scripps Research Institute calls a kind of 'geographic information system'—a human GIS.[203]

In 2007, Gary Wolf coined the phrase 'Quantified Self' in an article for *Wired* magazine.[204] Self-tracking using wearable sensors dates back to the 1970s, when athletes began recording their heart rates and the numbers of steps they ran each day. Now, monitoring devices can track everything from sleep to brain waves and include inventions like an electronic fork that helps you record the numbers of bites you take at a meal. Quantified Self conferences are held around the world and converts blog prolifically about their data, their interpretations of them and themselves, and their experiences trying to bring about personal change. The point of having such scientific data is to actively use it.

One scientist at the forefront of the growing self-study movement is Larry Smarr of the University of California San Diego. Smarr was one of the pioneers of the Internet and as director of the California Institute for Information Technology and Telecommunications, his attention turned to the possibilities of combining networked sensors with genomics. His 2012 article 'Quantifying your body' opens with the quote 'Quantifying oneself would have been only the stuff of dreams before the digital revolution'.[205]

Smarr is one of the best-studied humans on Earth, and talks and writes widely about his experiences. Like Eric Topol and other evangelists of personalized medicine, Smarr argues we must transform medicine with self-study. He prides himself in being an example of what 'systems medicine' might achieve. Leroy Hood, the father of systems biology, pioneered this concept, advocating what he termed the 'P4' approach to medicine. This is a paradigm shift towards 'predictive, preventative, personalized, and participatory' health-care practices.[206]

Smarr's initial motivation for self-tracking was to lose weight. His radical P4 journey started innocently enough; he aimed to lose 20 pounds after a move to sunny Southern California, the land of beautiful people, from the Midwest. Smarr tripled the number of steps he took each day. His REM periods of deep sleep, the most valuable periods of sleep, accounted for more than half the time he spent asleep, twice the typical proportion for a man of his age. His blood chemistry appeared in excellent working order. Yet he couldn't lose weight, despite his health and a strictly controlled diet.

When he spied high levels of the C-reactive protein (CRP) in his reams of data, a sign of acute inflammation, he finally was onto the culprit. He charged to his doctor, proud of his detective work, but came home disappointed. The medical profession struggles to keep up with the pace of research and is wary—with some reason—to jump on the latest bandwagon. Smarr lacked conventional symptoms; so nothing was done. 'Come back when you are sick—bring us symptoms' is a common retort that frustrates P4 medicine's proponents. Smarr was right to worry, however, and when he returned soon after with severe abdominal pains he was given antibiotics for an acute bout of diverticulitis.

The experience galvanized Smarr—he had diagnosed his problem before the medical establishment. He had more power than his doctors to understand his own health. He became fascinated with his gut microbes. He commenced more sophisticated self-tracking, DNA sequencing. He asked the questions core to human microbiomics. Who was there? Did he look 'normal'; what was 'normal'? Was his flora changing? What unintended side effects might the antibiotics have on his microbiome?

Smarr got to the root of his problem. He found the early stages of Crohn's disease. While not fatal, this condition brings with it a long list of painful symptoms that flare up from time to time. Detective work led him to realize his gut ecosystem was stressed; in fact it was undergoing a 'biodiversity collapse'. His protective 'good' bacteria had died back, making him more vulnerable to nefarious species. His medical history is not particularly unusual, but through his diligent self-study during the asymptomatic phase, he was one of the very few to see the train wreck coming.

Today, Smarr converses fluently about the relative abundances of different microbes in his system and how they are changing in response to his efforts to rebalance. Hearing him talk with other microbiome experts, one wonders if this might be hip cocktail party natter of the future. Due to his unique position at a leading research institution, Smarr was more easily empowered to take charge of his own health, but he adamantly wants everyone to have that opportunity. Medicine will catch up, thanks to champions like Smarr and Church. A genome sequence for all and an annual microbiome check are no longer viewed as the pipe dream of a handful of technologists.

Roller derby

Microbes are mobile. We carry a long-term store of microbes but we also drag and drop new organisms into our system as we pass through our lives. The way in which humans catch and trade pathogens—colds, coughs, flus, and worse—provides ample evidence that we trade microbes. Microbes readily move through our environments, whether via physical touch, the air, water, or other media like drinking fountains and toilets. Many infectious diseases originate in other animals: swine flu from pigs and HIV from primates—microbes also jump species boundaries.

Eric Alm and Lawrence David of MIT carried out a detailed study of the microbiomes of two healthy men over the course of a year. The men largely kept the same gut microbiota that they started with, but two events disturbed the peace. On both occasions the community shifted considerably but bounced back. One disruption was caused by a two-month trip to South East Asia and the other by a case of food poisoning. It shows how easily we pick up microbes from around us, most notably on the food we eat, but that a healthy system is resilient: it can recover from external shocks.[207]

Contact means swapping. A study led by Jessica Green at the University of Oregon looked at how non-pathogenic microbes move between humans. She chose a wonderfully unorthodox set of subjects: professional roller derby players.[208] Prior to competitions among teams from different cities, researchers swabbed the upper arms and shoulders of the skaters. Each team had its own brand of bacterial community. Like having uniforms of different colours, the microbes were just as good at grouping players into the right teams. When the women were swabbed after the game, each woman's skin had become more similar to the skins of members of the other team. It was as if players had mixed and matched parts of their uniforms or blended the colours or emblems into new 'recombinant' uniforms.

A growing number of studies are suggesting that such swapping is common—perhaps more so than might be comfortable for us. Spending time in a crowded room with many people, say at a three-day conference, means you will leave with a skin microbiome more similar to the 'crowd' and less like 'your own' one that you came with. Lotharios beware if your partner wants to compare belly-button microbiomes. The belly-buttons of long-term partners are more similar than people compared at random; sudden divergence with your partner might suggest an illicit dalliance.

We are in a constant dance with the microbes of our world, including the parts of it that we built. The study of the microbial dimension of the 'built environment' is a rapidly growing field. The questions are as rich as the patterns are fascinating.[209] It appears that we spread our microbial signature to each new place we inhabit. Doorknobs are some of the most microbe-rich places in your house. And yes, aeroplane bathroom walls are covered with aerosolized microbes. Some of us clutch bacterial sprays everywhere we go to ward off the invisible hordes. There are probably better ways to defend against the spread of pathogens. Some companies, for example, are developing sprays of 'good bacteria'. The concept is to apply these to countertops in hospital wards and other public places, maybe those aeroplane toilets, to help stave off the colonization and spread of 'bad' bacteria.

An epic study of 'microbial ebb and flow' is now under way. It focuses on one of our most special types of build environment—the clinic. The Hospital Microbiome Project[210] is a collaborative effort funded by the Sloan Foundation and led by Jack Gilbert of Argonne National Laboratory and the University of Chicago. The new hospital has arisen from its urban landscape like a volcano in the ocean, creating a new habitat for the city's microbes to colonize. A natural experiment at the landscape scale occurred in 1883 when Krakatoa erupted and a virgin island was formed. Scientists have learned a lot from studying the process of succession as waves of plants, insects, birds, and other animals colonized the island and turned it from barren rock to lush forest.

Gilbert's team is investigating the assembly of the hospital microbiome. They are tirelessly sampling rooms, equipment, patients, and staff.[211] The study is novel in its aim to understand the overall microbial ecosystem over time, not just the flow of a few pathogens through it when there is a disease outbreak. The first step is to understand the abundances, distributions, and movement of microbes within this special type of man-made island. In other words, it is a classic ecological study, but not one of plants and animals; this is microbial ecology. The beauty of the Hospital Microbiome Project is that it was launched before the building went up. Gilbert and colleagues will be able to observe how the hospital microbiome assembles and changes over time. Just like Smarr, they have established a new kind of observatory—a genomic observatory.

A buoy in the ocean

Another genomic observatory initiated by Gilbert sits out in the waters of the western English Channel 10 kilmometres off the coast of Plymouth in Devon. A port on the south-west coast of England, Plymouth has many claims to fame. Darwin departed from here on the 'Voyage of the *Beagle*' as did the Pilgrim Fathers 200 years earlier. It is also one of the longest studied marine sites in the world, with over a century of data collected from the Western Channel Observatory.[212]

Today an experimental buoy bobs up and down in the waves at the location of the Station L4, where the measurements of zooplankton, phytoplankton—the tiny plants and animals of the ocean—and myriad environmental factors are taken. Within some of the busiest shipping lanes in the world, many pass this unassuming little buoy but few know it marks one of the best-studied places in the world. Darwin would have passed very close by.

Striking patterns in microbial diversity were revealed here in metagenomic studies started by Gilbert, then at the Plymouth Marine Laboratory. Over a six-year time period, the microbial community found here is surprisingly predictable. Between 2003 and 2008, monthly seawater samples were collected from L4 for sequencing. The study generated 10,000 sequences per sample and revealed that the diversity of microbial species correlates strongly with day-length.[213] The longer the nights the more diversity is found. Diversity peaks on the shortest day of the year—the winter solstice—and dips on the summer solstice. Flushed by waters from the Gulf Stream every two weeks, the site possesses specific microbial communities depending on the season. Indeed, one can tell what month it is just from a profile of L4 microbes showing which are the most abundant.

Follow-up studies analysed the samples in more depth, and the patterns changed. The original data contained a huge number of rare species that were only seen once during the six years of sampling. One of the monthly samples was therefore sequenced again, but this time 10 million sequences were generated—this was 'deep sequencing'. Would more rare species be found? To the surprise of the researchers, the 'deep sequenced' sample contained all of the diversity found in the entire L4 time-series dataset—everything found in the 'shallow sequencing' of 72 samples, and then some. In fact, the six-year dataset accounted for only 4 per cent of the diversity found in this one deep-sequenced sample. This means that species composition stays throughout the year, but the relative proportions change. Some species might be more common in summer and others in winter, but they are always present.

To explore this further, the special sample was compared to ocean samples from around the globe. Then its importance became more obvious. The almost 400 buckets of water from around the globe amassed by the International Census of Marine Microbes, all looked quite different to each other. In other words, they showed the kinds of biogeographic pattern we typically expect from larger organisms. Australia has wallabies and eucalypts; Finland has reindeer and fir trees. They appeared to be structured geographically. But when compared to the single deep-sequenced L4 sample, much of the global diversity seemed to be in the English Channel too. It was as if we suddenly found a few koalas biding their time in Scotland waiting for the right conditions to arrive. For microbes, it seems that the adage that 'everything might be everywhere' could hold after all.[214]

One 2-litre bucket of water from the English Channel contained 44 per cent of the microbial species in the global study. If we continued to sequence water off the coast of Plymouth, would we eventually turn up all the marine microbial diversity on Earth? In fact, we can extrapolate to predict that it would require 200 billion sequences from Plymouth to find all known marine microbes. Could the entire ocean's microbial seed bank really be found off the coast of Plymouth? Of course, the caveat here is that we are comparing to all 'known' microbes. When more sites around the world are characterized as well as L4, the picture could change again.

We have learned an immense amount from L4. It demonstrates how we can learn about, possibly predict, an entire ocean by studying just one tiny part of it—one bucket of water—in unprecedented detail. The six-year study of marine microbes under the L4 buoy gave us a new view of sea life and how it responds to environmental change. Comparing the L4 time-series to samples from hundreds of other buckets of seawater gave us new insights into all our oceans.

Cottonwood, cod, and corals

The first species sequenced—the model organisms—were mainly those used in the laboratory for biomedical and agricultural research. We are now sequencing the genomes of wildlife. Of particular interest are the species that provide a foundation for entire ecosystems and that drive key ecosystem services. Ecosystem services are defined as services, or 'products', that humans derive from nature. Such services could be food, shelter, and other amenities of life including a range of biotechnological innovations as well as cultural and aesthetic values. It also covers such basic things as the pollination of crops by insects and the production of oxygen by microbes and plants.

Forests are among the most important ecosystems on Earth. We can use genomics to understand their ecological resilience and long-term sustainability. Tom Whitham and colleagues have led the way, focusing on cottonwood trees, poplars that include the model forestry species, black cottonwood, the first tree to have its whole genome sequenced.[215] Their research reveals that there is a lot of genomic potential even within a single species of tree. Just 100 metres of elevation can change the local environment enough to favour slightly different genotypes of cottonwood. This important genomic biodiversity remains hidden from view unless the trees are planted together in a common garden. When Whitham and others have done this, they find that the trees that grow best are those from the same elevation as the common garden. As little else differs, the performance of the trees depends mainly on their genomes. Common garden approaches have long been used in forestry. With the cottonwood genome in hand, Whitham and colleagues have been able to dissect the underlying mechanisms.

There is growing evidence that even single genes can influence the machinations of whole ecosystems. It is a huge leap forward to think of genes in an ecosystem working together. Whitham's team has pioneered the field of community genomics.[216] Their cottonwood research has shown that changes in genes can alter which herbivores and parasites feed on leaves. This in turn causes a cascade of changes that alters the entire ecosystem from the birds to the soil microbes. In other words, 'there are genetic components to understanding the response of plants to climate change, exotic invasions, and habitat restoration, and they can strongly interact'.[217] Whitham argues that even more important than the Endangered Species Act would be a 'Foundation Species Act'. While endangered species have their own intrinsic value, they are by definition now rare. Foundation species, on the other hand, are still common but that does not mean they are invulnerable. Losing a foundation species would result in massive ecological upheaval.

Whitham's work builds on a case made by Richard Dawkins in the early 1980s that genes can have an extended phenotype reaching well beyond the organisms in which they are embedded.[218] Genomics is now providing the capacity to observe these connections. We can see genes at work on the landscape scale. Genes are not visible from space, but microbes are when they mass together. Single-celled organisms living in the oceans produce some 50 per cent of our planetary

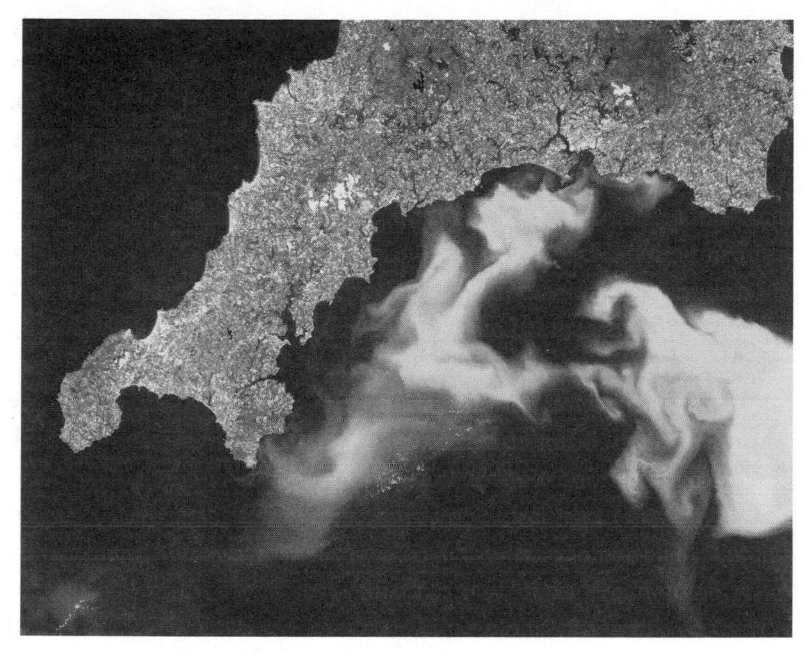

Figure 6. Genes visible from space. Blooms of calcium encased single celled *Coccolithophores* off the coast of southern England.

oxygen. Among the most prolific are the Coccolithophores (see Figure 6).

These beautiful, calcium-encased cells lived in such abundance that over time their shells built the White Cliffs of Dover. Today, we can capture images of blooms of these organisms from space, for example spanning the southern coastline of the UK. What causes the blooms to disappear is equally interesting. In some cases, it seems that a giant virus is wiping them out. Once the infection starts, the bloom is extinguished in short order. An epic battle of genomes is played out on a landscape scale. Now both the genome of *Emiliania huxleyi* and the giant virus that attacks it have both been sequenced in a bid to understand this phenomenon.[219]

Many plants and animals are part of the human food chain. We have already sequenced our domesticated species, such as cow, pig, and chicken, and now we are working on our wild sources of sustenance. The primary goal of 'food genomics' is the breeding of better stocks. Cod is a wild fish now being brought into domestication. It has been all but wiped out in its natural habitat. In general, domesticated species are replacing wild ones around the world. It is an example of human terra-genoming, although there is another theory: perhaps cows evolved to taste good so that they could enslave humans and take over the planet? If that is merely a joke, it is not so funny when one considers the amount of species lost as biodiverse forests are replaced with biologically bland pasture.

The loss of species is significant but perhaps even more important is the loss of genetic variation within species: genomic erosion. An insidious trend has taken hold. The rewards of artificial selection are going to only the tiniest fraction of genes and the most successful genomes. If the richest 1 per cent are getting richer in modern societies, while the remaining 99 per cent see their incomes squeezed, a similar phenomenon might be happening to the planetary genome. Some genomes are acquiring all the resources. Billions of genetic variants are being obliterated; bovine world domination approaches.

Domestication often results in the loss of genetic diversity. One side effect of engineered breeding can be that successful varieties dominate and others disappear. By selecting only certain breeds and specific cultivars we risk becoming dependent on an ever-narrowing slice of the total genomic diversity available to us. This can maximize yields in the short term but might weaken the resilience of our planetary life support systems. For example, pathogens and parasites are able to home in on abundant and static targets, threatening food security around the world.

Fish is a primary source of protein worldwide. The Atlantic cod provides an example of a wild, economically important species whose genome was sequenced expressly to help with domestication. We have removed 90 per cent of all big fish, the top predators, from our oceans in the last 50 years. Catching a full-grown wild cod is now almost a thing of the past. In 2003, the North Sea was closed for cod fishing. We are increasingly looking to marine aquaculture, the farming of seafood, as a solution. Cod, however, are notoriously difficult to farm. They are able to chew through nets and in aquaculture they are vulnerable to pathogens such as *Francisella noatunensis.* Given its considerable fishery and aquaculture interest, Norway sponsored the complete genome sequencing of cod.[220] At just under 1 billion base pairs the cod genome has some 25,000 genes. Researchers are exploring the genome for clues to the adaptability of cod. How does it respond to different temperatures and levels of oxygen, and what factors influence its growth and maturation?

The cod genome held some surprises. All vertebrates have what is known as a major histocompatibility complex (MHC): regions in the genome made up of hyper-variable genes that produce genetic variation to defend against pathogen attacks. What was thought to be a universal system, however, turns out to be not so universal after all. Compared to other vertebrates, the cod genome displayed 10 times as many genes in the MHC I region and a complete absence of genes in the MHC II region. The reasons for this deviation from the vertebrate norm are yet to be understood, but the discovery could help reduce disease incidence in aquaculture.

Farming can lighten the pressure on remaining wild stocks, but it brings its own woes. Green groups fear that escapees will contaminate wild populations, as farmed fish are prone to more diseases and parasites. Furthermore, a large influx of genetically homogeneous fish into the wild population could reduce overall genetic variation. Monocultures are known to be risky. One of the most infamous examples of this was the Irish potato famine. Farms relied on a single genotype of potato, which, while productive, offered no resistance against the new potato blight. Fear of such outcomes is driving the creation of seed banks to safeguard agricultural genetic diversity and the potential to develop new traits in future crops. We hold the 'gene pool' underlying human food supply in our hands, but we have already let much of it slip through our fingers.

Cod is a valuable species; it does us the service of being good to eat. Corals are a whole group of species that provide us with a whole range of ecosystem services worth billions a year. Coral reefs have been described as 'nature's ultimate jigsaw' and the 'rainforests of the sea'. They represent perhaps the greatest challenge in understanding the complex web that binds life together. Indeed at the heart of a coral reef lies a remarkable symbiosis. The animal host, the coral polyp, provides shelter within its cells for eukaryotic endosymbionts: dinoflagellate protists of the genus *Symbiodinium*. The coral polyp can forage for itself, but with the friendly *Symbiodinium* inside, it can capture the vast benefits of photosynthesis too.

Trees harness photosynthesis to build a forest, providing infrastructure for a multitude of other organisms. Their success also lies in the ancient collaboration between a eukaryotic cell and a prokaryotic bacterium that would eventually become the chloroplast. Two eukaryotes succeeded in the same trick, and corals use the energy they derive from photosynthesis to help lay down a skeleton. Like trees in a forest, corals provide habitat for all the other species on the reef. Indeed, corals go one better than trees: they even create their own land to live on. As young corals grow on top of the previous generation, their ancestors literally form the bedrock of the future. Coral atolls can persist for millions of years because they are living islands; they can grow to keep up when sea levels rise—as long as they don't rise too fast.

Coral reefs provide much of the best marine real estate in the tropics and support a huge and diverse wildlife. For humans, coral reefs provide subsistence and commercial fisheries, tourism and recreation, coastal protection from storms, a source of new bioactive compounds, and a host of cultural values.

The sequencing of a coral genome and its symbiont marks a shift away from the focus on sequencing model organisms towards a focus on communities of species that underpin key ecosystem services. In 2011, scientists of Japan's Okinawa Institute of Science and Technology sequenced the first coral genome. The genome of *Acropora digitifera* is relatively small at only 430 million bases and contains 23,700 genes.[221] In 2013, the first *Symbiodinium* genome was sequenced.[222] At 1.5 GB the selected reference species has a genome half the size of the human genome. It encodes some 42,000 protein-coding genes—double the number of humans.

Analysis of the coral genome revealed that corals evolved earlier than could be seen in the fossil record of hard-bodied corals. Reef-building Scleractinia first appeared in the fossil record in the mid-Triassic approximately 240 million years ago, but they were already highly diversified, suggesting much earlier origins, perhaps as soft corals that did not fossilize well. Comparison of sea anemone, hydra, and coral genomes suggests that these lineages diverged between 520 and 490 million years ago in the late Cambrian or early Ordovician. That is some 250 million years before the appearance of the first reef-building corals in the fossil record.

Corals, like humans, are microbial creatures. The reef ecosystem depends on coral and corals depend on microbes. Corals bleach white under stress. 'Ghost corals' are unattractive to the eye compared to the healthy glow of colours usually on show. It is a sign of sickness and even death. Coral bleaching occurs when their life-giving partners flee. As corals pale, much of the extended reef community suffers too. The main cause of coral bleaching seems to be heat stress, but there are other contributing factors too, including poor water quality. Deciphering the first genome of a coral symbiont will hopefully lead to a better understanding of coral bleaching.

The range of stressors impacting reefs is growing. Overfishing can be devastating because corals depend on other species to help them keep algae at bay. Algae compete with young corals for space on the reef and herbivores, like fish or sea urchins, help the coral win out. Destruction of fish nursery habitats or overfishing can tip the balance in favour of the algae. Greenhouse gas emissions are driving environmental changes that might exacerbate these local problems. Charles Keeley of the Scripps Institution of Oceanography raised the alert about rising carbon dioxide concentrations through his long-term atmospheric monitoring above Hawaii's high mountains.[223] The oceans also absorb a lot of the CO_2 we emit, and as CO_2 forms a weak acid in water, the average pH of seawater is gradually falling, a phenomenon known as 'ocean acidification'.

The icecaps are melting, but might the reefs melt away too? There is concern that ocean acidification will impact the ability of many marine animals, not least corals, to produce their hard skeletons. Will corals be able to grow fast enough in the new conditions to keep up with rising sea levels? That depends on the physiological tolerance of coral species, and to some extent upon genomics. Is there sufficient variation in the coral genomes to contain solutions? Might some corals be able to tolerate, even prosper, in the new environmental conditions? Or is there a threshold that no coral can survive and that we might just push them past?

Genomic biodiversity within foundation species, whether they are corals or trees, can increase the resilience of whole ecosystems. Whitham's group in Arizona, for example, has shown that some individuals of the pinyon pine tree are more vulnerable to insect pests. Why should susceptible genotypes persist under such selective disadvantage? The answer reveals why long-term studies are needed. When a drought hit the pine forest, the trees that survived best were those most vulnerable to the insects. It seems that drought resistance is under genetic control. Without sufficient genomic biodiversity, without this genetic potential of the pine population to resist insect pests and droughts, the whole forest community might be at risk.[224]

A similar situation seems likely in marine environments too. In corals, for example, genomic variation enables some corals and their symbionts to prosper in conditions that others find unbearable. For example, Daniel Barshis and colleagues compared corals that were sensitive to environmental stress compared to those that were more robust. They found that, when stressed, 'sensitive and resilient corals change expression of hundreds of genes, but the resilient corals had higher expression under control conditions across 60 of these genes'.[225] Such research offers novel solutions to climate change, including 'human-assisted evolution'. Amanda Mascarelli, writing in *Nature*, described the research as aiming to build designer reefs: 'creating resistant corals in controlled nurseries and planting them in areas that have been—or will be—hard-hit by changing conditions'.[226]

The Moorea Biocode

National Geographic photographer David Liittschwager started with a modest goal of exploring life in a single cubic foot of habitat over the course of a day. He wanted to make biodiversity more real to people and did so through vibrant imagery. Liittschwager is a pioneer of biodiversity art—and he found scientific soulmates in the Moorea Biocode Project. Collaborating with scientists from the Smithsonian, University of California Berkeley, and elsewhere, Liittschwager has developed a compelling concept that is both art and science: the 'biocube'.

His elegantly simple invention is constructed from bright green bars of metal, 12 inches on each side, the volume of 1 cubic foot. The frame he intentionally left open so life can travel through freely. A biocube is a standardized unit of biodiversity: pieces of habitat that would fit comfortably on your lap.

Liittschwager travelled around the world to produce a lavishly illustrated coffee table book.[227] Each chapter culminates in a collage of all the species found at a particular location. Liittschwager placed the biocube in the surface waters under the Golden Gate Bridge in San Francisco—his 'own backyard', as he calls it. In this coastal marine site, he photographed 83 species ranging from tiny diatoms to a Pacific harbour seal pup. This was just slightly fewer than the 97 species he found in Tennessee's Duck River, a recognized hot spot of freshwater biodiversity. The Duck River's bounty does not surprise, as it is a safe place for wildlife, surrounded by a relatively undeveloped rural landscape. The same cannot be said about the biocube Liittschwager sampled in Central Park, New York City. Yet even the heart of the world's greatest metropolis can be remarkably alive. Liittschwager

Figure 7. The Biocube anchored on the Temae reef in Moorea.

photographed 107 species in the deciduous forest of Central Park's Hallet Nature Sanctuary.

Travelling south of the equator and across the Atlantic, Liittschwager placed another biocube in the mountain Fynbos vegetation of Table Mountain National Park in South Africa and catalogued 113 species. In Costa Rica, at the Monteverde Cloud Forest Reserve, Liittschwager secured the biocube high in the branches of a tree and found 145 species. When he attached his biocube to the coral reef in the lagoon at Temae on the north-east corner of the South Pacific island of Moorea, he found 366 species—more than double the number of any other biocube (see Figures 7 and 8).

No study of biodiversity would be complete without a coral reef. At 17 degrees south of the equator, the ocean around Tahiti and the Society Islands has a natural lack of nutrients, which limits the growth of the microscopic phytoplankton that form the base of the pelagic food chain. In the bright tropical sunlight, the visibility around the Polynesian islands is therefore fantastic, making the sudden towers of life all the more shocking—mirages of biodiversity in an empty blue desert.

Liittschwager's coral reef biocube held 366 species, and with his scientific collaborators, he sequenced one gene—the DNA barcode—from each of them. But there are over 600 species of fish alone on the Moorea reef. Can we ever hope to sequence all the organisms on something as complex as an entire tropical island? The Moorea Biocode Project was launched in 2008 with the vision of doing just that—

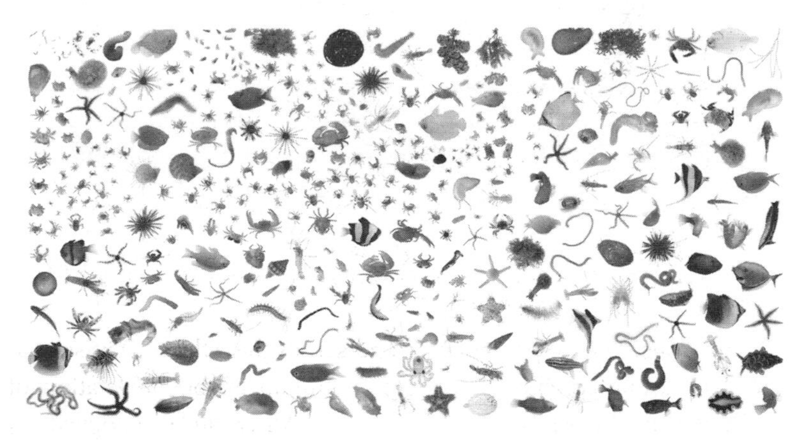

Figure 8. The Moorea biocube showing the 366 species seen on the reef.

at least for one gene per species to start. The project aimed to DNA-barcode the entire island, including all visible non-microbial life >1 millimetre in size. Having the DNA names of all species in this ecosystem would open up untold opportunities to study ecological processes.

In the tropical heat, organized by project director Chris Meyer from the Smithsonian Institution, researchers swam, dived, hiked, dug, and trapped.[228] Researchers from all over the world, each an expert on a particular taxonomic group, came to Moorea to comb its marine, freshwater, and terrestrial habitats. They scoured the nooks and crannies of its reefs and volcanic peaks for all the species they could find in their particular group.

At the start of the project, 5,000 species were known. As the project progressed, numbers grew. By 2012, the species list included more than 42,000 specimens from 6,580 species sampled during >4,400 unique collecting events along with >25,000 documentary photos. Now more than 8,000 species of animals and plants are estimated to live on the island and this still excludes fungi and invisible life. The known biological universe of Moorea continues to expand.

Around the world, researchers are now trawling in the invisible spectrum of life. Collecting adult specimens gives one view of the lagoon and ocean waters but is laborious and time-consuming. An alternative is to take a scoop of water and look at the small things. Marine waters are chock full of the young, or larvae. This includes the brood of everything from fish to sea urchins to corals. Much of the visible life in the waters around Moorea comes from invisible life. To capture this life nets are dragged behind boats at different depths and repeatedly in time. Everything caught is brought to the surface and sorted, using microscopes. Studies of the microbial life on the island are also advancing rapidly.[229]

Model me an island. The long-term aim of the study of Moorea as a model ecosystem is to build an avatar, or a digital representation of the island—an island in a computer. This would eventually be a computational model of place, including all DNA. This essentially attempts to fast-forward a century to a future when we have 'datafied' entire social-ecological systems and have sophisticated computer models to guide our development decisions.[230] This is the goal of P4 medicine at the scale of our bodies, our inner ecosystems. The Moorea idea is to apply this approach to our outer ecosystems too: the places in which we live—P4 sustainability.

This project is not a one- or two-year effort, but easily a 20-year venture as time, new technologies, and a huge integrative effort will be necessary to make headway. Creating the avatar will involve combining data on the physical and biological attributes of the island, from temperature, winds, and rainfall to patterns of waves and abundances and type of species—including humans—and looking for predictable patterns over time.

No sizeable ecosystem today can be fully understood without also taking into account social—human—attributes. Moorea is a particularly interesting, albeit complex 'model place', exactly because it has a sizeable human population of 17,000. The study of 'socio-ecological' systems, ones in which humans are considered part of nature, is at the forefront of sustainability science. The challenge now is to understand how all life interacts and to predict the consequences of these interactions, so many of which are characterized by complex feedback loops.

A key part of the avatar will be the inclusion of maps of biological interactions. One of the enabling conditions of the Moorea Avatar Project is the reference database created in the Moorea Biocode Project. Researchers are beginning to capitalize on this biological treasure trove, a biological map of the island, to do a range of studies. With the biological 'parts list' in hand, researchers can move to the next stage, following a roadmap developed by systems biology, to start to quantify and model how those parts interact to form the whole.

The food web is a core component of the island's 'interactome'. Understanding the Moorea 'who eats whom' is at the heart of building an avatar. It is nigh impossible in the field to observe all the possible meals consumed by *one* species, much less all of them. Imagine how hard someone, even a very close relative, would have to work to record everything you ate in a day, or a week. Now DNA analysis can be used to reconstruct the past. For fungi-eating beetles 'plate walks' can be used. A gluey substance covering the plate sucks everything off the tips of a beetle's legs as it tramps across—including fungi. Some beetle legs appear 'clean' and seed no fungal blooms. Such beetles fasted or dined on species too fastidious to grow on laboratory medium. Fungal cultures that grow to visible size are selected for DNA preparation and barcoding for identification.

For larger organisms stomach contents are DNA-barcoded. In one study led by Chris Meyer and Mathieu Leray from the French marine lab on Moorea, the CRIOBE, it was found that >95 per cent of the diet of the arc-eye hawkfish was recorded in the Biocode database but only <25 per cent of the related flame hawkfish. The two species showed little overlap in preferences and the new species found are presumed to be from transient or deep-sea species whose adult forms are not present on the reef where humans sampled. The best bio-collectors are not human.

The vision of the Moorea Avatar Project[231] is to integrate enough data to allow predictions to be made over the next 10–50 years but as the saying goes, in the end we are all dead, and it is clear what will happen to Moorea eventually. Several million years from now, Moorea will be gone; it will slip beneath the sea because of the natural process of erosion and subsidence of its volcanic peak. It is part of the Earth system and not independent of the larger processes around it.

Moorea is a remote island in the deep Pacific, yet its fate is linked to the rest of the globe. While local phenomena, like overfishing or pollution from the land, need to be addressed, the fate of Moorea will more likely be determined by events far away and global processes that no one on Moorea can hope to control. The biggest threat to the coral reef, which sustains the tourism and fisheries of the island, might come from higher ocean temperatures, rising sea levels, and ocean acidification. Smokestacks and car exhausts around the world are having a direct impact on the very heart of the Moorea ecosystem, its corals, and their microbes. No ecosystem is immune from the effects of industrialization. Even corals in the middle of the vast Pacific

Ocean are affected by human activities in the heart of Asia, Europe, or North America. Coral reefs are predicted to be the first major ecosystem on Earth to be lost if our current trajectory of human-induced environmental change continues. They are like canaries in the mine; unwilling subjects in an accidental geo-engineering experiment.

GEMs

On the evening of 19 July 2013, the day before the anniversary of the Moon Landing on 20 July 1969, NASA encouraged space enthusiasts to 'wave at Saturn'. The Cassini spacecraft snapped a picture of Earth, a distant speck visible between the glorious rings of Saturn at a distance of 1.44 billion kilometres.

Today, we are used to looking at the Earth as a system but this certainly wasn't always so, and many of us forget how interlinked life on the globe really is. A single photo triggered this epic change in human perception of our place in the universe—the 'Earthrise' photo, the most iconic photograph in history. Taken on Christmas Eve 1968 by the Apollo 8 mission, it was the first time we'd seen the Earth as a planet, rising above the barren landscape of another planetary body, the Moon.

The Earth, the blue planet, looks different from other planets. It is not only the anthropomorphic delusion that makes the Earth different—when we look at photos of the Earth from space, we can sense it is alive compared to similar images of other planets. Of one thing we can be sure, when we look at the Earth we are seeing the outward expression of one remarkable molecule—DNA. The Earth would not be blue-green with an oxygen atmosphere if it were not for the planetary genome.

The Earth is an island hanging in space. There is only one ocean, for example; plastics from our streets wash down into rivers and out into the oceans and collect in the ocean gyres. It is estimated that water cycles through this single system from rain to ocean floor every thousand years. Plastics are now found in the stomachs of deep-sea fish. Pregnant women and the most pristine reaches of the globe can be contaminated with traces of pesticide.

DNA circulates the planet, moving, mixing, and multiplying. The Earthrise photo is our first portrait of the planetary genome and all she has wrought. Can we ever model all life at the genomic level? Can we see the Earth's software and how it works? Efforts are at least starting to build planetary-scale ecological models of all life.

Drew Purves and colleagues at Microsoft Research and the United Nations Environment Programme World Conservation Monitoring Centre (both in Cambridge, UK) have led the call for a suite of 'Global Ecosystem Models' (GEMs). In their 2013 paper 'Ecosystems: time to model all life on Earth'[232] they point out: 'Obviously, modelling every organism within an ecosystem is impossible. (We estimate that it would take a standard laptop computer around 47 billion years to model for 100 years every multicellular animal within just one of the 1-degree grid cells covering Earth.)'

There are 360 longitudinal lines around the equator from east to west and 180 latitudinal lines running from the North to South Pole making 64,800 grid squares in total. To model the whole Earth would therefore take more than 3 trillion years of compute time. The age of the Earth is only 4.5 billion years while the Universe is only 13.8 billion years old.

So is there a hope of understanding this complexity? First, such models should be run on supercomputers not laptops, but Purves's illustration highlights the complex interactions that take place at the individual level. To overcome computational issues, his team makes simplifying assumptions, for example that all fish in a shoal behave in a similar fashion. So far, preliminary outputs from Purves's 'Madingley' model are broadly consistent with current understanding of ecosystems.

General Circulation Models, or GCMs, have profoundly helped us understand climate change. The Intergovernmental Panel on Climate Change (IPCC) shared the Nobel Peace Prize with Al Gore, the 45th Vice-President of the United States under Clinton, for 'efforts to build up and disseminate greater knowledge about man-made climate change, and to lay the foundations for the measures that are needed to counteract such change'. Now attention is rightly also focusing on life. The Intergovernmental Platform on Biodiversity and Ecosystem Services (IPBES), a sister of the IPCC, was established in Apil 2012 as a result.[233]

Society needs GEMs to do for biodiversity what GCMs have done for climate. We have worldwide infrastructure for measuring factors that produce the weather, from air temperature and speed to levels of humidity and rainfall. These types of data drive the generation of GCMs. Now we need more data on life.

Will genomics eventually feed up into such models? Genomic Observatories, sites of long-term research, are adding genomics to their infrastructures.[234] This emerging network is part of the far larger Group on Earth Observations (GEO), an international, decadal effort to build and unify existing infrastructures into the largest coordinated Earth-observing platform ever conceived. Integration of data from climate, weather, society and biodiversity, including DNA-level observations, will enable the vision of 'gene to satellite' flow of information into whole earth models. One pilot project of the Genomic Observatories Network is a global study of marine microbes, Ocean Sampling Day. Such work towards elucidating the function of DNA within complex systems is helping bring about a genomic synthesis, across scales and systems, planet down and genome up.

8

BIOCODING THE EARTH

The Biocode

Life on Earth runs off a shared biological code. The DNA of every organism is related, and combined, it represents a single entity: the Biocode. Descended from the first genome, the Earth Biocode has diversified and persisted for billions of years. It has shaped the world—right up to us.

Today, for the first time, DNA is not only fashioned by nature, but also written by man. DNA synthesizers are adding new pieces to the Biocode, fresh buds on the venerable 'tree of life'. The significance of this new-found human capacity remains to be seen, but it certainly marks a radical break in a multibillion year process. That has to be quite a big deal.

Genomics is mainly still in 'read-only mode'. Scientists around the world are sequencing genomes to learn how the software of life works. As we grow to understand the Earth Biocode, this knowledge will hopefully temper our age-old predilection to tinker blindly with the planetary operating system. At least it should give us further pause for thought.

In thinking about the future potential of genomics, however, gene sequences are far from the end of the story. Indeed, alone they are meaningless. Cells are what contain the information needed to execute genomic instructions, and despite the radical advances of synthetic biology, we cannot yet reproduce this cellular machinery artificially. An artificial cell is still a research goal.

The complexity of cells and developmental process remains a formidable scientific challenge that should not be underestimated, but progress is being made. For example, in 2012 a team of scientists led by Markus Covert of Stanford University announced the first computational model of an entire cell.[235] If we can build cellular 'avatars' *in silico*, it seems that truly synthetic life, a new tree of life, with both genomes and cells built by humans, from scratch, is probably only a matter of time.

A biocode can be defined as two or more interacting genomes. Dolly the Sheep had three mothers. It is conceivable that some human children might also have multiple mothers one day. Indeed, it is not uncommon for a surrogate mother to carry an embryo to term for the biological mother and such a baby could already have a third 'mother' if the child's mitochondrial DNA were donated to avoid inherited mitochondrial diseases.[236] Add the possibility of using a sperm donor, and a baby today could have five 'parents'.

We appreciate that no genome operates in isolation and a key goal of the future will be to unravel the relationships between genes in different genomes within the Biocode. We are just starting to see the mechanics of inter-specific, indirect genetic effects that shape communities and ecosystems.[237] The Moorea Biocode, all the island's interacting genomes, forms a web that extends beyond its tranquil lagoons. Yet threats of global change aside, things seem to work well on the island; everything seems to have its place. This is perhaps no mere myth of Polynesian paradise. Given enough time and stability, 'progress' might be the natural order of things. Cooperative systems (of genes) could tend to win out and as Richard Dawkins once put it: 'the world . . . a big island . . . tends to become populated by mutually compatible sets of successful replicators [genes], replicators that get on well together. In principle this applies to genes in different gene pools, different species, classes, phyla and kingdoms.'[238]

Increasingly, we are moving from sequencing one genome at a time to groups of genomes. In the future, we might all have our complete biocode sequenced (all our genomes). A biocode could be the genomes of any organism, a bioreactor, a building, or a larger system, such as the island of Moorea.

The 'biocode of' something need not imply that the entity is a functional unit. When all the organisms in David Liittschwager's cubic foot project were DNA-barcoded it was a glimpse of the biocode of his biocube, a standardized sample of life in space and time. His other biocubes contained their own miniature worlds, each with a unique set of genomes and organisms at a unique point in time: the biocube under the Golden Gate Bridge was a snapshot of the San Francisco Bay Biocode, and the one in Central Park, a snapshot of the New York City Biocode.

The 'L4 biocode' of the ocean, close to where Darwin and Venter started their epic voyages, is an even more open system. There seem to be no barriers to movement across the boundaries of the bucket-sized samples. It exists in the sense that scientists chose to sample that particular volume of water at that time. Interestingly, though, it showed the predictable hallmarks of a functioning system.

The intended flexibility of the word 'biocode' might make it hard to pin down scientifically; some might prefer 'metagenome', but it has the rigorous definition as being 'a group of genomes of related biological significance'. Key words in science are often notoriously difficult to define to everyone's satisfaction. If you have time to kill, ask a biologist for the definition of 'gene' or 'species'. Such words perhaps gain power through their relative vagueness. A biocode can be sampled at any scale or level of complexity.

Biocoding slips off the tongue, unlike 'DNA-ing' or 'metagenoming'. With it, we simultaneously acknowledge the huge diversity of currently 'independent' fields that all focus on understanding DNA and its consequences, all rightly having specialized and precisely defined scientific terms to describe what they do. Genomics (and the related fields of microbiomics, metagenomics, and DNA barcoding) is stimulating the proliferation of a growing number of subfields that span the biomedical and ecological sciences. They include personal, synthetic, evolutionary, ecological, and environmental genomics. New on the horizon is biodiversity genomics—sustainability genomics is an example of a new 'omics' about to crest. A grand unification is under way and biocoding is at the heart of the genomic synthesis that is defining the coming decade—and this century of biology. In biology, DNA is the emerging lingua franca and the chronicles of the century of biology will be written in it.

Our place in nature

We remain overwhelmed by the enormity of the Earth Biocode. We have yet to see most of it. Miescher initially wanted to be a priest, but went to medical school instead. Shyness and deafness made it impossible for him to deal with patients and he preferred the laboratory. It was there that he discovered 'nuclein'. He was surprised to find it not only in pus washed off bandages but also in salmon. Many have had the sense that all life is connected. Darwin provided the most convincing theory for how that might indeed be the case. Miescher was the first to physically see what made it so: DNA. Thanks to Mendel, Watson, Crick, and many others, we now know the underlying mechanism that unifies life, confirming Darwin's great insight.

Genomics is to life what the Periodic Table is to chemistry. 'Nothing in biology makes sense except in the light of evolution' was the title of a classic paper written by Theodosius Dobzhansky.[239] As a geneticist, Dobzhansky recognized the key contribution of DNA to understanding evolution and explaining the diversity and unity of life. But genomics cannot answer all the questions we have about living organisms or the ecosystems they form. Just as knowing about individual elements cannot explain all the properties of complex molecules, knowing genomes does not explain everything about the organisms and ecosystems that they form. Many phenomena are 'emergent'; they happen at higher levels of organization. The internal combustion engine does not explain all the maddening behaviour of traffic. Although answers to questions in the life sciences do not end with DNA—they start there.

Today's instance of the Earth Biocode, the planetary genome, includes all of us and every other living organism. From the tiny acorn of LUCA, the first cell, and MILO, its first genome, a vast tree of life has burgeoned to dominate the planet. Over billions of years, it has formed soil, filled the atmosphere with oxygen, and—acting through humans—it has built great cities.

There is still so much DNA left to explore. The vast majority of species and genes are waiting to be discovered. They encode solutions to the challenges faced by living organisms over the aeons of Earth history, including times when the planet had drastically different conditions. The planetary genome represents a treasure trove of parts that will spur the industrialization of synthetic biology. The ingenuity of life is contained in this genomic heritage; it is natural genomic capital.

Genomics is adding exquisite layers of detail to our knowledge of life and its relationships. The branch of the Tree of Life that holds Darwin's beloved pigeons, for example, a model of artificial selection, has recently exploded. Darwin surmised that all domesticated breeds came from the rock dove. In 2013, the public nucleotide repositories went from holding almost no sequences from pigeons to containing 38 complete genomes. This data showed conclusively that rock doves are indeed ancestral to the more than 1,200 breeds of fancy pigeons that exist—more than any other plant or animal in human history.[240]

The fact that all pigeons, and all living organisms, are related through descent, gives us a huge foot up in understanding the biology of all life. While this unity is encouraging, the sheer size of the Tree inspires awe. We know there are millions of species not yet formally described and more to be discovered. There are more microbial cells on the planet than stars in the Universe. They represent 50 per cent of the biomass and 99 per cent of the genetic diversity on Earth.

We are but twigs. Phylogenetic analyses, coupled with archaeological, anthropological, and other kinds of data, show that modern humans arose in Africa a mere 200,000 years ago—one of the most recent species to emerge from a procession of life that has been running for more than 3.5 billion years. We sit at the very tips of a Tree that is more like an exploding star, filled with deeply branching single-celled Protist lineages and an equally bewildering array of Bacteria and Archaea—most yet undocumented. Parts of the tree are actually more like a web because of the movement of genes 'horizontally' among sometimes-distant lineages, especially between microbes. We may someday need to add a fourth domain: alien life from another planet or synthetic life we create here on Earth.

Sunjammer

In late 2014, scientific visionary, author, and lifelong proponent of space travel Arthur C. Clarke will have started an epic voyage into deep space. More precisely, Clarke's DNA, found in a few donated strands of hair, will be making the voyage. His DNA will travel 3 million kilometres towards the sun aboard the Sunjammer, a solar wind sail powered craft launched by Celestis.[241] This is a most modern way of honouring the passing of loved ones—sending their cremated remains on 'memorial spaceflights'. In genomics, science fiction rapidly seems to become reality. Who better to epitomize this trend and the opportunities at hand than Arthur C. Clarke?

The gaps are still huge, but we will continue to fill in our digital record of the planetary genome. Genomics projects are inflating as quickly as technologies and ambitious, curious minds envision them. Nick Loman has a fun blog post that lists single publications with the most genomes. Entitled 'The biggest genome sequencing projects: the uber-list!', he is tracking the biggest projects.[242] The list is quite telling. At the time of writing, it is dominated by studies of bacterial pathogens, a few of which generated not hundreds but thousands of genomes. Currently, top of the list is a paper reporting 3,615 genomes of group A *Streptococcus*, the 'flesh-eating' pathogen.[243] Other studies include 3,000 rice genomes and 2,007 *C. elegans* genomes. Yet more large-scale projects spanning many species, like the '100K Foodborne Pathogen Genome Project', also continue to be announced.

Increasingly, Fortune 500 companies, like IBM, are starting to care about genes—in particular, microbial biodiversity. In May 2014, IBM hosted the conference 'Sequence the City: Metagenomics in the Era of Big Data'.[244] CNET reported: 'Part of a project known as the Almaden Institute, the work is part of a large effort to spread the study of "microbiomes" beyond medicine to other industries, including agriculture, food safety, counter-terrorism, forestry, forensics, retail, public utilities, and others.' Interviewed by CNET, James Kaufman, a manager at the IBM Almaden Research Center, is quoted as anticipating that 'with the costs of sequencing dropping, it will be routine to sequence anything and everything'. This could include buildings and even whole cities.

We are moving well beyond our penchant for sequencing human genomes, laboratory model organisms, and our pathogens. We are starting to tackle all of life; this is where 'biodiversity genomics' really begins. The first institute to bear this name was launched in Canada. The Centre for Biodiversity Genomics (CBG) opened in 2013 to the tune of 50 million Canadian dollars.[245] An extension of the Biodiversity Institute of Ontario (BIO), which opened in 2007, it is led by Paul Hebert, founder of DNA barcoding for species identification. He wants to DNA-barcode the world. Beyond whole tropical islands, like Moorea, entire countries are now getting in on the action, such as the German Barcode of Life.

Just as the Quantified Self movement is building a toolkit and infrastructure for self-study, we are beginning to quantify the planet with appropriate instruments and devices and these efforts are just starting to think about DNA. Might we someday have genomic sensors we wear? The X Prize Foundation has already offered a prize for 'a portable, wireless device in the palm of your hand that monitors and diagnoses your health conditions'.[246]

The P4 approach, espoused by Larry Smarr's microbiome self-study, is also spreading to a new 'health sector', the one concerned with maintaining good environmental status. What makes a healthy ecosystem? Can we diagnose and cure ills at the landscape level? Best of all, can we prevent sicknesses in the first place? Every place, like every person, is unique but shares some characteristics with other places. Prediction of future environmental problems and preventing them from happening in the first place can save economic pain and social suffering. Ensuring the wellness of our ecosystems requires the same kind of active management advocated by P4 medicine—what might be called 'P4 sustainability'.

We have a long history of environmental sensing for weather and climate[247] and now prototype DNA-detectors are being developed. Chris Scholin of the Monterey Bay Aquarium Research Institute (MBARI) is developing ecogenomic sensors and applying them to the study of ocean microbes, the base of much of our food chain, as a step towards documenting the Earth's DNA and understanding its function. Studies of the waters off the coast of California and Hawaii found evidence that bacteria can orchestrate metabolism across species.[248] In such a network of species, the waste of one organism is the food of another and 'community-level' metabolism can realize complex chemical pathways that cannot be contained in one genome. This provides further evidence of the extraordinary linkages between genes and the organisms that possess them.

Can we one day hope to measure biodiversity as we measure temperature or the concentration of carbon dioxide—or our own temperatures? Biodiversity is hugely variable and its many forms resist standardized measurements. One aspect of biodiversity, however, is consistent, measurable, and universal: the strings of A, C, G, and T that constitute the biocode. We cannot read DNA from satellites, so worldwide coverage will require *in situ* genomic sensing. There are increasing efforts at national levels to ensure such long-term study is carried out methodically and consistently. Funded to the tune of 0.5 billion US dollars over 30 years, for example, the US National Ecological Observatory Network (NEON) includes routine DNA barcoding of various organisms and metagenomic analysis of soils across the entire country.[249]

Why might we need a global view of DNA? One answer is because gene interactions recognize no political boundaries. They are connected through a vast network, a system that extends from each genome up to the planet. There are potential patterns in this genetic system, the planetary genome, which might only be visible from a global perspective. Are we losing genetic diversity as a planet? Only standardized longitudinal dataset across the planet can provide an answer. We can see greenhouse gas concentrations rising worldwide; might we be able to detect genomic diversity eroding, or the spread of antibiotic resistance, or the emergence of new infectious diseases? A global genomic observatory could aggregate a multitude of *in situ* observations to address such questions.

You too can biocode

Genomics reminds us we are not separate from nature. In the footsteps of naturalist explorers, like Darwin, comes a modern cadre of 'DNA explorers', such as Eric Karsenti of the European Molecular Biology Laboratory in Heidelberg, Germany. Voyaging the world in the yacht *Tara*, Karsenti and the other scientists of 'Tara Oceans' plied the seven seas imaging and sequencing plankton, from bacteria to protists.[250] But you don't need a yacht to be a DNA explorer. Nor do you need the access to sophisticated laboratories of pioneers like Larry Smarr or Michal Snyder. There are more and more opportunities for anyone to get involved—as citizen scientists.

Citizen science is a growing phenomenon in which the general public can take part in scientific experiments and help to document the natural world. Citizen science projects educate and raise awareness about genomics; they can also bring samples to light that change our view of the world. The National Geographic Society, like other venerable institutions, is reinventing itself in the digital era, and has become a pioneer in DNA exploration. In 2005, the society partnered with IBM to launch the Genographic Project.[251] Led by explorer-in-residence

Spencer Wells, Genographic aims to recount 'the greatest story ever told': that of prehistoric human migrations.

Genographic engaged members of the public, especially indigenous peoples. Anyone else could participate simply by purchasing a kit and submitting their DNA for analysis. Genographic has now genotyped more than half a million people. One day, into this vast pool of data, came a completely novel Y chromosome. When it 'failed' Genographic's standard suite of analyses, it was passed along to the DNA ancestry company Family Tree DNA for more extensive study. Sequencing revealed so many ancient mutations compared to modern Y chromosome lineages that it is estimated to be up to 338,000 years old.[252] Such a deep age significantly pre-dates Mitochondrial Eve (up to 200,000 years old) and previous estimates of Y Adam (estimated at 60,000 years old). Subsequently, this ancient Y chromosome has been found in a number of men within the Mbo people of Cameroon.

Members of the public can also peek at their microbiome. The company uBiome[253] and the research project American Gut used the Indiegogo crowd-funding platform to develop campaigns for public microbiome studies.[254] In 2013, uBiome offered a range of self-analysis options under the modest invitation to 'Learn about your health & change the world!' For US$5 you could support uBiome. U$25 would get you a company T-shirt and the promise of making you a 'Science Fashionista'. U$79 bought a gut microbiome test and with higher purchases one could sample additional locations including mouth, nose, skin, genitals, and add more than one time point. For U$10,000, one could visit the uBiome lab and get a full expert consultation.

uBiome more than tripled their goal of US$100,000, raising $351,000. Participants received kits with instructions that read: 'avoid bathing or bringing any substances that might disturb the microbiome into contact with the sample site for at least eight hours before sampling. This includes contact with antiseptic or antibiotic soaps or lotions, sex, kissing, food, hot tubs and pools, and the like.' Following the success of the campaign, accompanied by a storm of media coverage, uBiome now runs a formal for-pay service with tests starting at $89.

Part of the larger Human Food Project, the American Gut is a research project—an open community making all its data public. It raised even more funds than uBiome—£405,000 from 2,705 contributors—but the totals climbed afterwards to more than £700,000 from some 9,000 contributors. As co-founder of the Human Food Project, Jeff Leach is driven by an interest in understanding the anthropology of microbes and has written a book about it, entitled *Honor Thy Symbionts.*[255] As an anthropologist, he is working to understand how our relationship with food has changed over the millennia. For example, he is studying some of the last true hunter-gatherers on Earth, Tanzania's Hadzabe people, to understand how they form associations with their microbes.

Such projects are so powerful because they build on existing stores of data to provide essential interpretive context. The American Gut website illustrates the power of this combined approach using data from the Human Microbiome Project. For example, it can be shown that microbial communities across humans differ in the mouths, skin, gut, and vagina. A newborn baby gut community looks like a vaginal sample, reflecting the mechanism of delivery. Over a two-year period it changes in composition to look like an adult gut. If you generate your own data in American Gut, you can compare it with guts around the globe.

Biodiversity is not just in our gut but on it too; it turns out we have a whole world in our navel. Biologist and writer Rob Dunn, another pioneer of citizen science, is happy to make good science fun. He is famous for publishing a 2012 study of 'belly-button' microbiomes.[256] After swabbing the belly buttons of hundreds of volunteers and sequencing the samples he could say that, on average, humans host 67 species of microbes in that special place—a number similar to skin in general. Carl Zimmer, the well-known science writer, proudly advertised he took part.[257]

While it was hard to predict the overall composition of navel microbes on any one person, as this varies considerably between people, there seemed to be a group of microbial 'oligarchs' that dominate bellybutton-land. As in a human oligarchy, the power appears to be held by a small number. Only six types of critters, or phylotypes, dominated in >80 per cent of humans. While these accounted for less than 1 per cent of the total species, they made up to one-third of the cells found in any one belly button.

More recently, Dunn launched the 'Wildlife of Our Homes Project'.[258] His survey of more than 1,000 citizen science samples from houses promises to build an 'atlas of house-associated diversity', whether bacteria, fungi, or insects. Today just about anyone can set up a genomic telescope and train it where they please, be it their navel, gut, home—or beyond.

The Planetary Genome Project

Only a century before Friedrich Miescher discovered DNA in 1869, Captain Cook observed the Transit of Venus in Tahiti. At the time, navigating there was almost as impressive as the first Moon Landing two centuries later. On 5 June 1769, Cook and other scientists around the world took part in the first globally coordinated 'Big Science' project in history; improving our estimation of the distance from the Earth to the Sun—the so-called Astronomical Unit—by observing the Transit of Venus.

At that time, telescopes and clocks were cutting-edge technologies and Europe was in her grand Age of Discovery. Still, scientists travelled, in organized fashion, to the furthest flung locations on Earth to collectively watch Venus pass in front of the sun and use the combined data to understand a fundamental truth about our Solar System.

Only three centuries later, on 20 July 1969, we had taken the double helix to the Moon—or perhaps it had taken us there. Whichever the case, the long reach of the gene, as Richard Dawkins characterized it in the *Extended Phenotype*, finally extended beyond our planet. Today DNA sequencing technology is radically overhauling our understanding of life and we are moving into the age of 'actionable genomics', taking action based on genomic information. In only three centuries, we may be irretrievably into the Earth's sixth mass extinction: the Anthropocene extinction.

We are of little geological significance in the life of our planet, but these six centuries of technological advancement, starting from Cook's first voyage to Tahiti, promise to be the most transformative in human history and—with impending synthetic life and mass extinction—biological history.

Cook sent Joseph Banks, the ship's naturalist, from Tahiti to nearby Moorea to observe the Transit of Venus. To watch the great event, Banks officially set up the first scientific observatory on Moorea. By 2012, the 300-strong Moorea Biocode Project team had completed the first genetic catalogue of all visible life on the island and was beginning to conceive of the Moorea Avatar initiative to model the entire ecosystem, genomes and all.

On 5 June 2012, a team of researchers, led by us, observed the Transit of Venus from the very same spot where Banks stood, to symbolically launch the first genomic observatory, the Moorea Genomic Observatory. We also took the first water samples of microbes to launch a pilot run of Ocean Sampling Day (OSD), the first coordinated action of the Genomic Observatories Network.

OSD involves the standardized sampling and sequencing of a unit of seawater. Modelled on the work carried out at the L4 buoy in the Western Channel Observatory, OSD aimed to coordinate such monitoring globally. Two years later, the main June 2014 OSD event became the first simultaneous, coordinated megasequencing project, involving a large number of sites from the Genomic Observatories Network across the world.

OSD gives a glimpse of what could emerge as a single global genomic observatory: a DNA observatory designed to study the Earth Biocode, from individual organisms, including each human, right up to the planetary scale. OSD 2014 assembled parts of such an instrument, albeit in the most rudimentary form. It involved the collaboration of researchers at 180 marine sampling sites; while coordination was provided through a European Union research grant, participation in the event was voluntary, showing the power of collaborative science. OSD also included a citizen science effort focused on measuring environmental parameters of ocean waters on the day, for example, involving sailors from the 'Summer Sailstice'.

The 500 or so OSD collaborators sent out boats, and researchers took samples for metagenomic analysis and environmental measurements. DNA samples from OSD will go to the Smithsonian's Global Genome Initiative for bioarchiving; part of the Global Genome Biodiversity Network of museums. This kind of spidering out of science, combining core central facilities, like museums and genomics laboratories—usually found in cities—with a global network of field stations—some very remote—is the vision of future studies of the planetary genome.

Banks—and Cook—would have been astounded to learn that every organism, including all of us, had its name written inside it, and that this name was the entire recipe for building the organism. Like Cook and Banks, Charles Darwin also contemplated the view from Tahiti. His gaze was not focused on the stars, however, but on the coral reefs of Moorea. Entire atolls were formed by life, corals laying down skeletons thanks to the ability of their microbial symbionts to harvest sunlight—all part of the Biocode. Darwin would have enjoyed knowing that DNA sequencing would help prove the living world was one big family, and that both his theories of evolution—of atolls and life—would stand the test of time.[259]

Here we come to the conclusion of this book. The British astronomer Edmund Halley predicted the 1761 and 1769 transits of Venus almost 40 years before they occurred and recognized the opportunity to better estimate the Astronomical Unit. Knowing he would not live to observe them, he issued a call to action that inspired international scientific effort to observe the rare event.[260] His plan required the action of governments, funding bodies, scientific societies, and

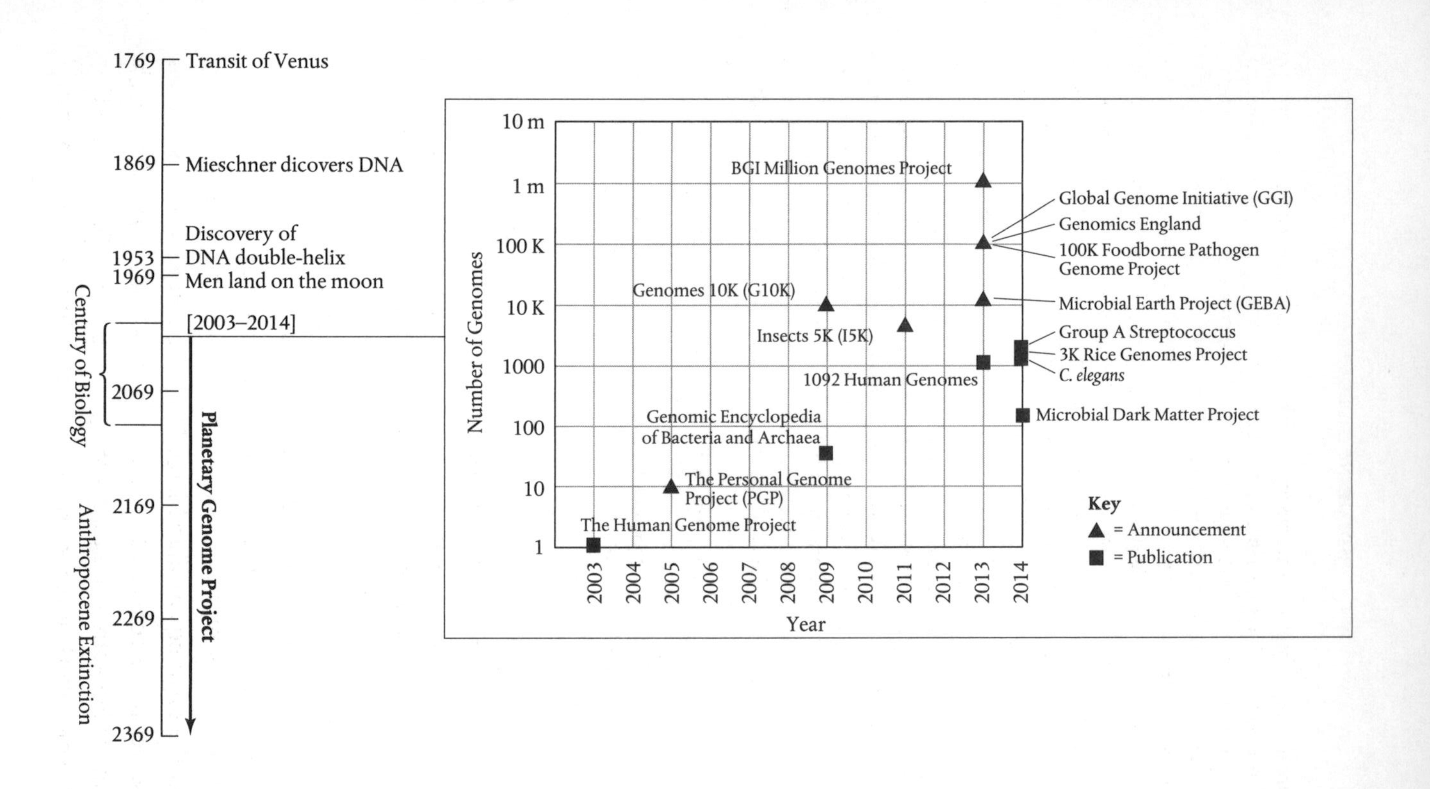

1769
Transit of Venus
1869
Mieschner dicovers DNA
1953
Discovery of
DNA double-helix
1969
Men land on the moon
[2003–2014]
2069
Century of Biology
2169
2269
2369
Anthropocene Extinction
Planetary Genome Project
Number of Genomes
10 m
1 m
100 K
10 K
1000
100
10
1
Year
2003
2004
2005
2006
2007
2008
2009
2010
2011
2012
2013
2014
BGI Million Genomes Project
Global Genome Initiative (GGI)
Genomics England
100K Foodborne Pathogen
Genome Project
Genomes 10K (G10K)
Microbial Earth Project (GEBA)
Insects 5K (I5K)
Group A Streptococcus
3K Rice Genomes Project
C. elegans
1092 Human Genomes
Microbial Dark Matter Project
Genomic Encyclopedia
of Bacteria and Archaea
The Personal Genome
Project (PGP)
The Human Genome Project
Key
= Announcement
= Publication

Figure 9. Six centuries of change—from the Transit of Venus to the Anthropocene extinction. This is the century of biology and genomics is at its heart. Born in 1995 with the sequencing of the first bacteria genome, genomics went meteoric in 2003 with the sequencing of the human genome. Coming only 50 years after the discovery of the DNA helix, it took only little more than a decade for the first publication describing >1,000 human genomes to be published and a call to be made for sequencing 1 million genomes. Growth in the size and number of genomics projects continues exponentially, but of course publications of actual genomes (triangles) lag behind more ambitious project announcements (squares). Key projects described in this book are plotted using year and number of genomes to show this trend. Publications with thousands of genomes from a single species started appearing in 2013. Microbiomic and metagenomic projects are also rapidly escalating in scope, from the Sargasso Sea study (1 site) to the Global Ocean Survey (41 published of 200 sampled sites) to Ocean Sampling Day (+180 sites) to the Earth Microbiome Project (hundreds and growing). Still, these ambitious projects only cover a small slice of total genomic biodiversity. Catching up with the total number of species on Earth will take time. The iBOL project already contains some 300,000 DNA names of species but estimates of the total number of species on Earth will certainly far exceed 10 million when microbes are finally catalogued. Compared to today's largest genome sequencing projects, a proposal to sequence all the DNA-barcoded organisms on the island of Moorea (8,000) would now be considered a 'smallish megasequencing project'. The combination of all of these projects, and the many more certain to be conceived, form the de facto Planetary Genome Project.

hundreds of scientists and their citizen science colleagues to complete. They travelled the world, pooled their data, and came up with a good estimate of the Astronomical Unit. It helped humanity to understand its place in the Universe and paved the way for the exploration of space.

We have another major scientific goal in sight now—understanding the software that shapes our living planet. A de facto Planetary Genome Project is under way to understand the Earth's Biocode, from single genomes to pools of genome to the planet. It is the sum of all the ongoing small and large DNA sequencing projects and all those yet to be conceived. Increasingly, it is being realized through highly collaborative, megasequencing projects, like Ocean Sampling Day, which work at the international level to leverage global research infrastructures, but will most importantly be the nexus of searches into every corner and crevice for clues about how genomes work to create life and its consequences.

Spurred on by the 'Big Science' of the Human Genome Project, we are now sequencing the rest of life. Like Halley's dream of observing the Transit of Venus, this grass-roots Planetary Genome Project will run to the four corners of the Earth and advance to the degree that data is shared, pooled, and interpreted. Barely five decades from the elucidation of the structure of DNA, and just one decade since sequencing the human genome, we are embarking on biocoding the planet.

ENDNOTES

1. James Watson and Francis Crick's classic paper was the first to describe the double helical structure of DNA. With some understatement they note that the structure 'suggests a possible copying mechanism for the genetic material' (Watson & Crick, 1953).
2. Dawkins described DNA as an immortal coil in his book *The Selfish Gene* (Dawkins, 1976).
3. Dalí's view on DNA is described in *The Molecular Gaze: Art in the Genetic Age* (Anker & Nelkin, 2003), one of the first books about the influx of DNA into art.
4. On the 50th anniversary of the publication of the DNA double helix, *The Economist* discussed the push of DNA into popular culture, quoting Soraya de Chadarevian in an article entitled 'The art of DNA: back to bases' (<http://www.economist.com/node/1730781>).
5. Church & Regis's book describes the field of synthetic biology and sets out a vision for the future (Church & Regis, 2012).
6. 'Base pairs' refers to the nucleotide molecules—the 'letters' A, C, T, G—that are linked together to make up a strand of DNA. Sequencing machines can 'read' the order in which these letters occur and the resulting sequence is what GenBank stores. The numbers describing the growth of public DNA sequence data are taken from the annual release notes for the world's public DNA databases held in the US, Europe, and Japan: NCBI-GenBank Flat File Release 198.0 (<ftp://ftp.ncbi.nih.gov/genbank/gbrel.txt>).
7. The US National Center for Biotechnology Information hosts the world's public repository of DNA data known as GenBank. It is apt that it is a part of the larger National Library of Medicine (NLM) run by the US National Institute of Health (NIH). GenBank is part of the International Nucleotide Sequence Database Consortium (INSDC) which includes mirror databases at the European Bioinformatics Institute (EBI) in Cambridge, UK and the DNA Database of Japan (DDBJ).
8. This topic was discussed in a paper on the compression of sequence information to aid storage. Public stores of DNA are growing faster than available hard-disk space (Cochrane, Cook, & Birney, 2012).
9. The estimate of the total digital information on Earth comes from an article by Martin Hilbert and Priscila López (Hilbert & López, 2011) and the comparison to the number of galaxies was made in a news story covering their paper: <http://phys.org/news/2011-02-world-scientists-total-technological-capacity.html>.

10. Goldman and Birney's work on DNA storage (Goldman et al., 2013) was widely covered in the mass media.
11. Ewan Birney makes a tongue-in-cheek call for a billionaire philanthropist to help make a 10,000-year DNA archive and outlines what would have to be solved to do so on his blog 'Genomeinformatician': <http://genomeinformatician.blogspot.co.uk/2013/01/the-10000-year-archive.html>.
12. For a description of PCR and other educational material related to DNA, see Cold Spring Harbor Laboratory's DNA Learning Center website <http://www.dnalc.org/>. The section on PCR is found at <http://www.dnalc.org/resources/animations/pcr.html>.
13. Good coverage of this unbelievable story of false fatherhood can be found on CeCe Moore's website 'Your Genetic Genealogist' and starts with this post: <http://www.yourgeneticgenealogist.com/2014/01/artificial-insemination.html>.
14. The Health Street website is here: <http://www.health-street.net/dna-tests.html>. The 'Who's Your Daddy?' truck has been widely covered in the popular press. It hit the streets first in New York, then Boston, and is aiming for other major US cities.
15. The VGL's website contains information on the DNA tests available for different animals (<http://www.vgl.ucdavis.edu>).
16. See 'New policy at Midtown Crossing tests dog DNA' which describes one of the first communities to monitor the DNA of all its dogs: <http://m.ketv.com/news/new-policy-at-midtown-crossing-tests-dog-dna/-/17419034/21845976/-/82gyjwz/-/index.html>.
17. The contentious Supreme Court decision on the use of DNA in routine arrests was widely covered in the mass media, such as a story in the *New York Times* entitled 'Justices allow DNA collection after an arrest' (<http://www.nytimes.com/2013/06/04/us/supreme-court-says-police-can-take-dna-samples.html?_r=0>).
18. The story of 'Tinker' the cat was widely covered in the mass media, including an articled entitled 'Cat DNA database used to catch killer' in the *Huffington Post* (<http://www.huffingtonpost.co.uk/2013/08/14/cat-dna-database-uk_n_3753138.html>).
19. This DNA artist's website is <http://deweyhagborg.com/> and includes links to her 'Stranger Visions' project of DNA portraits created from human trash picked up on the streets of New York City.
20. The Genspace lab opened in 2009 and strives to promote citizen science and access to biotechnology: <http://www.genspace.org/>.
21. The *New York Times* reported on 'Do-it-yourself genetic engineering' on 10 February 2010, featuring MIT's International Genetically Engineered Machine (iGEM) competition (<http://www.nytimes.com/2010/02/14/magazine/14Biology-t.html?pagewanted=all&_r=0>). The democratizing of molecular technologies is encouraged through initiatives like 'DIY Biology' (<http://

DIYbio.org>) founded in 2008. DIY DNA is also a phrase that appears in the tabloid press; for example, the *Daily Mail* reported on 20 January 2014 on the 'Soaring sales of "dangerous" do-it-yourself DNA test kits', see <http://www.dailymail.co.uk/news/article-2542956/Soaring-sales-dangerous-DNA-test-kits-Number-websites-selling-products-doubles-two-years.html#ixzz3AooDTtg2>.

22. See 'DNA prediction of human eye and hair colour from ancient and contemporary skeletal remains' (Draus-Barini et al., 2013). Software for 'Modeling 3D facial shape from DNA' (Claes et al., 2014) was reported in *Nature* as 'Mugshots built from DNA data' (Reardon, 2014).
23. A video of the founders describing the Miinome concept can be seen here: <https://angel.co/miinome>. The Miinome website has since been launched and currently takes your 23andMe data and offers further interpretation and access to a social community (<http://miinome.com/>).
24. The announcement of the human genome sequence proved one of the biggest science stories of the decade. The quote about the genome being a most wondrous map was widely covered. For example, see these articles from *ABC News* and the *New York Times*: <http://abcnews.go.com/Technology/story?id=99380&page=1> and <http://partners.nytimes.com/library/national/science/062700sci-genome-text.html>.
25. For an entertaining and informative account of the race to sequence the human genome, see Jamie Shreeve's book *The Genome War: How Craig Venter Tried to Capture the Code of Life and Save the World* (Shreeve, 2005).
26. The public and private sectors published back-to-back human genome papers in 2001 as the culmination of the race between Collins's and Venter's groups respectively: 'Initial sequencing and analysis of the human genome' (Lander et al., 2001) and 'The sequence of the human genome' (Venter et al., 2001).
27. The 'Venter Genome' was published in an article entitled 'The diploid genome sequence of an individual human' (Levy et al., 2007).
28. Venter's book (Venter 2013) details the story of his creation of life through synthetic biology and considers what the definition of 'life' is, both historically and in the context of the genomics era.
29. The 'Watson Genome' was published in an article entitled 'The complete genome of an individual by massively parallel DNA sequencing' (Wheeler et al., 2008).
30. For a discussion of pharmacogenomics see the US National Library of Medicine website, which describes it as a 'new field combin[ing] pharmacology (the science of drugs) and genomics (the study of genes and their functions) to develop effective, safe medications and doses that will be tailored to a person's genetic makeup': <http://ghr.nlm.nih.gov/handbook/genomicresearch/pharmacogenomics>.
31. See the Scripps Translational Science Institute's 'Wellderly Genome Reference' website: <http://www.stsiweb.org/SWGR/>.

32. In 2006, US$10 million was on offer for the Archon Genomics XPRIZE but it was withdrawn once they realized that 'genome sequencing technology is plummeting in cost and increasing in speed independent of our competition'. See <http://genomics.xprize.org/>.
33. The Personal Genome Project continues to enrol volunteers and all data is on the public webpage here: <http://www.personalgenomes.org/>.
34. Church outlined his vision for the Personal Genome Project in this paper published in 2005 (Church, 2005).
35. Pinker's description of taking part in Church's PGP and having his genome sequenced is here: <http://www.nytimes.com/2009/01/11/magazine/11Genome-t.html?_r=0>.
36. Elizabeth Silverman wrote extensively about genomics as a biotechnology stock analyst at Punk Ziegel and Robertson Stephens. The article cited here is 'Genomics, biotechnology's oldest next big thing—FierceBiotech' (<http://www.fiercebiotech.com/story/genomics-biotechnologys-oldest-next-big-thing/2013-05-08#ixzz2VzfMJRX3>).
37. The homepage of the International HapMap Project is here: <http://hapmap.ncbi.nlm.nih.gov/>.
38. See the US National Library of Medicine website for a discussion of SNPs at <http://ghr.nlm.nih.gov/handbook/genomicresearch/snp>.
39. Complete Genomics has since been acquired by the BGI (<http://www.completegenomics.com/news-events/press-releases/archive/Complete-Genomics-Adds-29-High-Coverage-Complete-Human-Genome-Sequencing-Datasets-to-its-Public-Genomic-Repository--119298369.html>).
40. For the announcement of Denmark's FarGen project, see the project website at <http://www.fargen.fo/en/> and also the February 2013 report in *Bloomberg* entitled 'Faroes' 50,000 residents leap into DNA testing quagmire' (<http://www.bloomberg.com/news/2013-02-25/faroes-50-000-residents-leap-into-dna-testing-quagmire.html>).
41. See 'An integrated map of genetic variation from 1,092 human genomes' (Abecasis et al., 2012) and one of the follow-up papers interpreting the dataset (Khurana et al., 2013).
42. See an opinion piece on discussions between the FDA and 23andMe in the *New England Journal of Medicine* (Annas & Elias, 2014).
43. See a discussion of 'Variations in predicted risks in personal genome testing for common complex diseases' (Kalf et al., 2014).
44. The future of companies like 23andMe is still being determined. The first class action lawsuit filed against 23andMe in the Southern District Court of California is reported in *Forbes* magazine: <http://www.forbes.com/sites/danmunro/2013/12/02/class-action-law-suit-filed-against-23andme/>. See court document at <http://docs.google.com/file/d/0BozYPQn3U6APbjhyMUxvN2ZtVEU/edit>.

45. Cecile Janssens reported on 'How FDA and 23andMe dance around evidence that is not there' in her blog on *Huffington Post* <http://www.huffingtonpost.com/cecile-janssens/post_6753_b_4671077.html>.
46. The development of the Genomics England initiative (<http://www.genomicsengland.co.uk/>) was widely covered, for example, in the online journal *Genomeweb*: 'Genomics England to sequence 8K genomes for two pilots in 2014, establish sequencing centers' (<http://www.genomeweb.com/sequencing/genomics-england-sequence-8k-genomes-two-pilots-2014-establish-sequencing-center>).
47. The inclusion of genomics self-testing in training medical students appeared in the *Mercury News* article 'Stanford University students study their own DNA' (<http://www.mercurynews.com/science/ci_23310506/stanford-university-students-study-own-dna>). Information on the course Gene210 can be found here: <http://www.stanford.edu/class/gene210/>.
48. Angelina Jolie's article 'My medical choice' appeared in the *New York Times*. She chose to tell the story of her double mastectomy as a result of finding out her BRCA genotype in a bid to help other women and their families coping with hereditary breast cancer: <http://www.nytimes.com/2013/05/14/opinion/my-medical-choice.html? r=0>.
49. In this book, Eric Topol describes how new sensor technologies and genomic information is helping to drive the personalized health care revolution (Topol, 2013).
50. The annual *MIT Technology Review* of the 50 most disruptive companies in 2013 included the sequencing giant Illumina as well as BGI: <www2.technologyreview.com/tr50/2013/>.
51. See <http://www.technologyreview.com/featuredstory/524531/why-illumina-is-no-1/>.
52. The BGI's Million Genomes Project was announced in 2011 and a project webpage is here: <http://www.genomics.cn/en/navigation/show_navigation?nid=5658>.
53. Topol points out that at least 20,000 individuals are needed to identify rare, functional genomic variants. This, coupled with the capability to sequence a human genome for US$1,000 in 2014, will lead to a 'virtuous cycle of informativeness' as millions more will want to sequence themselves and the sequencing industry achieves ever greater economies of scale (Topol, 2014).
54. See the review 'Genome mosaicism—one human, multiple genomes' (Lupski, 2013).
55. For an example of somatic tissue mosaicism, see 'Extensive genetic variation in somatic human tissues' (O'Huallachain et al., 2012).
56. For an example of genomic mosaicism as the genetic basis for a range of neurological diseases, see the paper 'De novo germline' (Rivière et al., 2012).

57. The study of Y chromosomes in female brains has been repeated in other systems and the same high rates of mosaicism are found other tissues as well (Chan et al., 2012).
58. For an excellent and detailed review of the emerging field of epigenetics, see Nessa Carey's book *The Epigenetics Revolution: How Modern Biology Is Rewriting Our Understanding of Genetics, Disease, and Inheritance* (Carey, 2013).
59. The BabySeq project is one of several projects (Kaiser, 2013) now being funded to sequence the genomes of infant cohorts and follow the impact this has on their health; see <http://www.genomes2people.org/babyseqproject/>.
60. The source for how many 'genomic labels' are provided by the FDA is the article 'Individualized medicine from prewomb to tomb' (Topol, 2014).
61. See the UK National Health Service website: <http://www.nhs.uk/Conditions/Downs-syndrome/Pages/Introduction.aspx>.
62. Researchers presented what they claimed was 'the first step towards a potential genetic/epigenetic approach to chromosome therapy' (Jiang et al., 2013).
63. Cancer genomics is the biggest growth area in genomics today and efforts are widespread. The international ICGC is by far the largest single project and its homepage can be found here: <http://www.icgc.org/>.
64. 'No longer human' was the title of a piece by Lori Oliwenstein in the December 1992 issue of *Discover Magazine* (<http://discovermagazine.com/1992/dec/nolongerhuman171/>). Oliwenstein was reporting on the claim that HeLa cells are single-celled microbes that are closely related to humans, but their own distinct species (Van Valen & Maiorana, 1991).
65. HeLa is 'the most widely used model cell line for studying human cellular and molecular biology' (Landry et al., 2013).
66. See *The Immortal Life of Henrietta Lacks* (Skloot, 2010).
67. Skloot's opinions on the publication of the HeLa genome with consent of the Lacks family can be found in the *New York Times*: <http://www.nytimes.com/2013/03/24/opinion/sunday/the-immortal-life-of-henrietta-lacks-the-sequel.html?_r=0>.
68. On the creation of the HeLa Data Use Agreement, see <http://www.bio-itworld.com/2013/08/07/henrietta-compromise-nih-announces-hela-data-use-agreement.html>.
69. The decimation of the Tasmanian devil by infectious cancer has been widely covered in the media. One article in *Time* reports on the creation of a 'Devil's Ark' to aid in their long-term survival: <http://newsfeed.time.com/2013/01/23/noahs-ark-to-save-tasmanian-devils-from-cancer-plague/>. The Devil's Ark project can be read about here: <http://www.devilark.com.au>.
70. Cited 50,000 times, this is one of the most important papers in the field of biology and helped forge the field of bioinformatics. 'BLAST' stands for 'basic local alignment search tool' and allows fast comparisons of the similarities between DNA and protein sequences (Altschul et al., 1990).

71. This is one example of an increasing number of genomic studies designed to aid conservation and support population genomics, in this case of the Tasmanian devil (Miller et al., 2011).
72. Discussions of the technology and ethical consideration of de-extinction are being discussed in several arenas. For example, see the TedxDeExtinction conference in 2013 which profiles the views and work of a large number of leading academics working in genomics, synthetic biology, cloning, conservation, policy, and ethics: <http://www.ted.com/tedx/events/7650>.
73. Church's idea of bringing Neanderthals back into the world through a human mother surrogate was widely covered in 2013: <http://www.telegraph.co.uk/science/9814620/I-can-create-Neanderthal-baby-I-just-need-willing-woman.html>.
74. The Venter interview in the *New York Times* in which he discusses his hopes for creating novel life and the 'Hail Mary Genome' is found here: <http://www.nytimes.com/2012/06/03/magazine/craig-venters-bugs-might-save-the-world.html?pagewanted=all>.
75. See 'The minimal gene complement of Mycoplasma genitalium' (C. M. Fraser et al., 1995).
76. See 'Global transposon mutagenesis and a minimal Mycoplasma genome' (Hutchison III, 1999).
77. See 'Generating a synthetic genome by whole genome assembly: phiX174 bacteriophage from synthetic oligonucleotides' (H. O. Smith et al., 2003).
78. See 'Complete chemical synthesis, assembly, and cloning of a Mycoplasma genitalium genome' (Gibson et al., 2008).
79. See 'Genome transplantation in bacteria: changing one species to another' (Lartigue et al., 2007).
80. See 'Creation of a bacterial cell controlled by a chemically synthesized genome' (Gibson et al., 2010).
81. For a stimulating philosophical discussion of genetic versus non-DNA inheritance (i.e., the molecular machinery of the cell) and whether referring to genetic programmes is meaningful and useful or not, see the article by Denis Noble of Oxford University, entitled 'Genes and causation' (Noble, 2008).
82. *Wall Street Journal* interview with Craig Venter at Singularity University on 6 September 2013: <http://live.wsj.com/video/craig-venter-on-synthetic-life-genome-sequencing/D40D0F00-BFAD-48F7-8857-08ABABEE439A.html#!D40D0F00-BFAD-48F7-8857-08ABABEE439A>.
83. A call for a synthetic human genome was published in the *Huffington Post*: <http://www.huffingtonpost.com/andrew-hessel/human-genome_b_1345842.html>.
84. See an article reporting the event in the *New Scientist*: 'First baby born after full genetic screening of embryos' (<http://www.newscientist.com/article/dn23827-first-baby-born-after-full-genetic-screening-of-embryos.html?

full=true#.UxoyT-ewLIQ>). The results were first presented by Dr Dagan Wells of the NIHR Biomedical Research Centre at the University of Oxford, at a scientific meeting in London on 8 July 2013: see abstract at <http://www.eshre2013.eu/Media/Releases/Dagan-Wells.aspx>.

85. See the article on GenePeeks in *MIT Technology Review* (20 November 2012): 'Genetic screening can uncover risky matches at the sperm bank' (<http://www.technologyreview.com/news/507491/genetic-screening-can-uncover-risky-matches-at-the-sperm-bank/>).
86. 23andMe filed a patent for a future 'genomic simulator' for prospective parents that was covered, for example, by *Wired* magazine in this article: http://www.wired.com/wiredscience/2013/10/23andme-patent/>, patent number 8543339.
87. This work by Levenson and colleagues (Haase et al., 2013) was widely covered in the press, for example, in a *CBS News* article: 'A happy marriage may depend on your genes' <http://www.cbsnews.com/news/happy-marriage-may-depend-on-your-genes>.
88. Juan Enriguez is developing the concept that the intersection of technology and genomics is culminating in the emergence of a new human species, *Homo evolutis* (Gullans & Enriquez, 2011). See his TED Talk on the topic: <http://www.youtube.com/watch?v=JNcLKbJs3xk>.
89. In July 2014, there was much press coverage that a British company had developed an app that 'lets Google Glass wearers control the device using their mind alone'; see <http://www.telegraph.co.uk/technology/google/10958566/Google-Glass-gets-mind-control-app.html>.
90. Extensive information about work towards the 2045 human avatar concept is presented on the project webpage, including a detailed technical roadmap. A brief summary of the concept is presented in this online video: <http://motherboard.vice.com/blog/russian-billionaire-dmitry-itskov-plans-on-becoming-immortal-by-2045>.
91. After Sanger's success in producing the 'Nucleotide sequence of bacteriophage *Φ*X174 DNA' (Sanger et al., 1977), he and his colleagues skipped other animals to target directly the human mitochondrion. It was the largest molecule that had been sequenced to date with ~16k base pairs (Anderson et al., 1981).
92. 'Genome' was already being used by another journal at the time. Roderick recalls coming up with the word 'genomics' in an interview with Bob Kuska (Kuska, 1998). He was credited with the suggestion in the editorial of the first issue of *Genomics* (McKusick & Ruddle, 1987).
93. See 'Whole-genome random sequencing and assembly of *Haemophilus influenzae* Rd' (Fleischmann et al., 1995).
94. While *H. influenza* is forever smeared as the flu bug, the viruses actually responsible have never received a Latin name under the Linnaean system—the scientific standard for nomenclature. They belong to a still formally undescribed 'species' known as Influenza A virus, which has a number

of subtypes, named after the characteristics of two surface proteins: hemagglutinin (HA) and neuraminidase (NA). The virus that caused the 1918 pandemic is referred to as subtype H1N1. More recently, other subtypes have hit the headlines, such as bird flu caused by subtype H5N1.

95. Mojo Nixon's song 'Elvis is everywhere' can be heard at <http://www.youtube.com/watch?v=e_hkIN38qnY>.
96. See 'Toward automatic reconstruction of a highly resolved tree of life' (Ciccarelli et al., 2006).
97. See 'Complete genome sequence of the methanogenic archaeon, *Methanococcus jannaschii*' (Bult et al., 1996).
98. See yeast genome 'Life with 6000 genes' (Goffeau et al., 1996).
99. The Genomes Online Database (GOLD), <http://genomesonline.org/>, (Kyrpides, 1999) is the premier website cataloguing complete and draft genome and metagenome projects. Over 200 fields of information are collected and a range of statistics routinely included. The complete dataset can be downloaded in spreadsheet form. Regular publications appear in the journal *Nucleic Acid Research* (Pagani et al., 2012).
100. See 'Genome sequence of the nematode C. elegans: a platform for investigating biology' (C. elegans Sequencing Consortium, 1998).
101. See 'Analysis of the genome sequence of the flowering plant Arabidopsis thaliana' (Arabidopsis Genome Initiative, 2000).
102. See 'Whole-genome shotgun assembly and analysis of the genome of Fugu rubripes' (Aparicio et al., 2002).
103. In a classic paper, geneticist Theodosius Dobzhansky put it as follows: 'It is a matter of opinion, or of definition, whether viruses are considered living organisms or peculiar chemical substances. The fact that such differences of opinion can exist is in itself highly significant. It means that the borderline between living and inanimate matter is obliterated' (Dobzhansky, 1973).
104. The Giant Viruses website describes what is known about giant viruses in general starting from the sequencing of the famous 'mimivirus': <http://www.giantvirus.org/>. The subsequent discovery of the super-huge genomes of *Pandoravirus* was widely covered in the scientific and public media (Philippe et al., 2013; Yong, 2013).
105. See 'Small, smaller, smallest: the origins and evolution of ancient dual symbioses in a Phloem-feeding insect' (Bennett & Moran, 2013), and 'Genome degeneration and adaptation in a nascent stage of symbiosis' (Oakeson et al., 2014).
106. See 'The Amborella genome and the evolution of flowering plants' (Amborella, 2013).
107. See 'The Norway spruce genome sequence and conifer genome evolution' (Nystedt et al., 2013).
108. See 'Genome sizes through the ages' (Leitch, 2007) and the 'Animal Genome Size Database' at <http://www.genomesize.com>.

109. GeneSweep was widely covered with *Science News* publishing the story of Rowan's win on 3 June 2006: <http://news.sciencemag.org/2003/06/low-numbers-win-genesweep-pool>.
110. See 'Genomics in *C. elegans*: so many genes, such a little worm' (Hillier et al., 2005).
111. See 'Draft genome sequence of the sexually transmitted pathogen *Trichomonas vaginalis*' (Carlton et al., 2007).
112. Looking back on the first decade since the human genome was sequenced, 'the most important outcome of the human genome project has been to expose the fallacy that most genetic information is expressed as proteins' (Mattick, 2011). The relationship between non-protein-coding DNA and eukaryotic complexity, and the role of non-coding DNA has triggered a paradigm shift (Mattick, Taft, & Faulkner, 2010; Mattick, 2007; Taft, Pheasant, & Mattick, 2007).
113. The ENCODE Project's landmark paper, published in *Nature*, claims that the consortium managed 'to assign biochemical functions for 80% of the [human] genome, in particular outside of the well-studied protein-coding regions' (Bernstein et al., 2012).
114. See 'Genomic organization of human transcription initiation complexes' (Venters & Pugh, 2013).
115. See 'Genome analysis of multiple pathogenic isolates of *Streptococcus agalactiae*: implications for the microbial pan-genome' (Tettelin et al., 2005), the first paper to sequence enough strains from one species of bacterium (six) to recognize and describe the concept of a pan-genome. This paper launched studies of myriad other pan-genomes and the widespread recognition of this higher-level genetic 'organization' in bacteria.
116. Scott Edmund's 'Tweenome' blog post offers a good summary of the history of the event: <http://blogs.biomedcentral.com/gigablog/2011/08/03/notes-from-an-e-coli-tweenome-lessons-learned-from-our-first-data-doi/>.
117. All data remain available via BGI's *GigaScience* database under the most liberal public domain access rights: the Creative Commons Zero (CC0) 'no rights reserved' waiver. Creative Commons is one of the organizations driving the movement towards making creative content, including music, scientific papers, and any other intellectual creation, more readily accessible. The CC0 waiver means that the content, in this case the genomic data, can be used by anyone for any purpose without even any legal obligation to credit the source. This latter feature is perhaps the one that is hardest for scientists to give up, but not obliging a citation is not the same as forbidding it, and scientific etiquette and norms still require the crediting of sources. The CC0 is the same licence Church selected for his exceptionally open Personal Genome Project.
118. The *Science as an Open Enterprise* report can be downloaded from <http://royalsociety.org/policy/projects/science-public-enterprise/report/>.

119. See 'Metagenomics to paleogenomics: large-scale sequencing of mammoth DNA' (Poinar et al., 2006).
120. See 'A high-coverage genome sequence from an archaic Denisovan individual' (Meyer et al., 2012). The Denisovan Genome Consortium made the data available for download prior to publication: <http://www.eva.mpg.de/denisova>.
121. The publication of the Thistle Creek horse genome from fossil bones received a lot of press attention because it pushed back the date for extracting viable DNA from ~80,000 years to almost 700,000 years. See 'Recalibrating Equus evolution using the genome sequence of an early Middle Pleistocene horse' (Orlando et al., 2013).
122. This work was widely covered, for example by *Nature* who led their article with the statement that 'Genetic material can't be recovered from dinosaurs—but it lasts longer than thought' <http://www.nature.com/news/dna-has-a-521-year-half-life-1.11555>. See 'The half-life of DNA in bone: measuring decay kinetics in 158 dated fossils' (Allentoft et al., 2012).
123. See 'Animals in a bacterial world, a new imperative for the life sciences' (McFall-Ngai et al., 2013) for a discussion of the importance of bacterial genes in animal evolution. In their paper, McFall-Ngai et al. present the relative percentage of the human genome that arose very early in biological evolution, citing an earlier paper: 'An ancient evolutionary origin of genes associated with human genetic diseases' (Domazet-Loso & Tautz, 2008).
124. See 'The genome of the choanoflagellate *Monosiga brevicollis* and the origin of metazoans' (King et al., 2008).
125. See 'A phylogeny-driven genomic encyclopaedia of Bacteria and Archaea' (Wu et al., 2009).
126. See 'Genomic encyclopedia of Bacteria and Archaea: sequencing a myriad of type strains' (Kyrpides et al., 2014).
127. See 'Insights into the phylogeny and coding potential of microbial dark matter' (Rinke et al., 2013).
128. See 'Genome 10K: A Proposal to Obtain Whole-Genome Sequence for 10 000 Vertebrate Species' (G10K Community of Scientists, 2009) and the G10K website <http://genome10k.soe.ucsc.edu/>.
129. The sequencing of Darwin's finch was published as a dataset in the journal *Gigascience* (Zhang et al., 2012) and described in the paper 'Insights into the evolution of Darwin's finches from comparative analysis of the Geospiza magnirostris genome sequence' (Rands et al., 2013).
130. See 'Creating a buzz about insect genomes' (Robinson et al., 2011); the community made a call for the iK5 project and the project homepage: <http://www.arthropodgenomes.org/>.
131. For GIGA, see <http://nova.edu/ocean/giga/>.

132. Javier Del Campo and colleagues have called for an effort to generate better coverage of eukaryote species in their review article 'The others: our biased perspective of eukaryotic genomes' (Del Campo et al., 2014).
133. For GGI, see <http://www.mnh.si.edu/ggi/>.
134. The history of the formation of the Human Microbiome Project is described in this 'marker paper', which establishes the research community: 'The NIH Human Microbiome Project' (Peterson et al., 2009).
135. David Relman and Stanely Falkow made their call urging consideration for a 'second human genome project' in an article on 'The meaning and impact of the human genome sequence for microbiology' (Relman & Falkow, 2001). Relman's team went on to publish another highly influential paper in 2005 on the 'Diversity of the human intestinal microbial flora' where they used genetic techniques (ribosomal RNA gene sequencing) to demonstrate that there were many more microbial species in the human gut than could be found through classical cultivation methods. They concluded that 'Characterization of this immensely diverse ecosystem is the first step in elucidating its role in health and disease' (Eckburg et al., 2005).
136. The website <http://nihroadmap.nih.gov/hmp/> provides additional information about the HMP and access to its data.
137. Nelson's first metagenomic microbiome paper 'Metagenomic analysis of the human distal gut microbiome' (Gill et al., 2006) helped define this field, comparing two healthy adults, and concluding that 'humans are superorganisms whose metabolism represents an amalgamation of microbial and human attributes'. The first phase of the HMP went on to produce 'A catalog of reference genomes from the human microbiome' (Nelson et al., 2010).
138. The MetaHIT results are published in 'A human gut microbial gene catalogue established by metagenomic sequencing', finding a bacterial 'gene set, approximately 150 times larger than the human gene complement' (Qin et al., 2010).
139. Microbe pioneer Rob Knight explains what microbes do around the body and how they might cure disease at a TED talk in 2014 (see his blog <http://blog.ted.com/2014/03/19/how-microbes-could-cure-disease-rob-knight-at-ted2014/>) and in his book *Follow Your Gut: The Enormous Impact of Tiny Microbes* (Knight & Buhler, 2015).
140. See 'Microbial ecology: human gut microbes associated with obesity' (Ley et al., 2006) and a sister paper 'An obesity-associated gut microbiome with increased capacity for energy harvest' (Turnbaugh et al., 2006) showing that certain gut bacteria (in mice) do consume more energy, giving a mechanistic basis to a role in weight gain/loss.
141. See 'Gut microbiota from twins discordant for obesity modulate metabolism in mice' (Ridaura et al., 2013).
142. See 'Geographical variation of human gut microbial composition' (Suzuki & Worobey, 2014).

143. Michael Pollan describes his experience with the American Gut Project and cultivating his own microbiome as a result of the food he consumed and his lifestyle here: <http://www.nytimes.com/2013/05/19/magazine/say-hello-to-the-100-trillion-bacteria-that-make-up-your-microbiome.html?pagewanted=4&ref=magazine>.
144. Carl Zimmer's *New York Times* article 'Tending the body's microbial garden' (Zimmer, 2012) cites a study entitled 'Microbiota-targeted therapies: an ecological perspective' (Lemon et al., 2012) and recounts a paradigm shift away from the language of warfare when referring to microbes and the need for an ecological approach.
145. See the review paper 'Replenishing our defensive microbes' (Ursell et al., 2013) that expounds on new ways of thinking about microbes—through the ways they help us.
146. Costello and colleagues argued that the human microbiome provides 'health-related ecosystem services' when they considered 'The application of ecological theory toward an understanding of the human microbiome' (E. K. Costello et al., 2012). They drew explicit parallels between microbial processes and classical community ecology (of plants and animals); in particular, how the development of the microbiota in infants represents community assembly in previously unoccupied habitats; how recovery from antibiotics represents assembly after disturbance events; and how pathogens represent the impact of invasive species.
147. These plans are described further in this news article: <http://www.news-medical.net/news/20140513/Navy-asks-Rice-synthetic-biologist-to-tweak-gut-bacteria-for-mood-weight-control.aspx>.
148. The concept that microbes might go extinct is increasingly being explored. For example, the article 'Bugs inside: what happens when the microbes that keep us healthy disappear?' appeared in *Scientific American* as early as 2009: <http://www.scientificamerican.com/article/human-microbiome-change/>. More recently, the focus is on how the impact of widespread use of antibiotics on the human microbiome could be responsible for many of today's emerging diseases. An excellent review of this research can be found in Martin Blaser's book *Missing Microbes: How the Overuse of Antibiotics Is Fueling Our Modern Plagues* (Blaser, 2014).
149. Marie Myung-Ok Lee's article is at <http://opinionator.blogs.nytimes.com/2013/07/06/why-i-donated-my-stool/?_php=true&_type=blogs&_r=0>.
150. Speaking to *Nature* magazine in June 2013, Gary Wu, a gastroenterologist at the University of Pennsylvania in Philadelphia, pointed out that 'Stool is a very complex mixture that we don't fully understand' (Mole, 2013). So there you have it. We still don't know shit. But we are learning fast. In 2012, Alm and Mark Smith established OpenBiome, which maintains a stool bank: 'Samples are homogenized, filtered and frozen for long-term storage, providing physicians with a standardized, convenient source of material.'

It is cool shit. Eventually, Alm and others envisage that we will learn enough about the 'active ingredients' in the stool transplants to develop synthetic communities for a more targeted next generation of microbiome therapeutics (M. B. Smith, Kelly, & Alm, 2014).

151. BGI published the iconic panda genome 'The sequence and *de novo* assembly of the giant panda genome' (Li et al., 2010) as part of their efforts to raise the profile of genomic sequencing in China.
152. When genes for digesting bamboo weren't found in the panda genome, researchers looked at the panda's microbes and found 'Evidence of cellulose metabolism by the giant panda gut microbiome' (Zhu et al., 2011).
153. This genome sequencing of 34 pandas is expected to aid conservation efforts: 'Whole-genome sequencing of giant pandas provides insights into demographic history and local adaptation' (Zhao et al., 2013).
154. The story of hunting for biofuel genes in panda gut microbes is here: <http://news.nationalgeographic.com/news/energy/2013/09/130910-panda-poop-might-help-turn-plants-into-fuel>.
155. 'The draft genome of the fast-growing non-timber forest species moso bamboo (Phyllostachys heterocycla)' shows that bamboo has a genome of 2.05 Gb and 31,987 genes (Peng et al., 2013). Apart from pandas, about 2.5 billion people depend economically on bamboo, and international trade in bamboo amounts to over US$2.5 billion per year.
156. Fierer and colleagues, in their study on 'Reconstructing the microbial diversity and function of pre-agricultural tallgrass prairie soils in the United States', showed how cultivation has significantly changed the microbial composition of soils in the American Midwest (Fierer et al., 2013).
157. Bias in the molecular primers used in previous studies of soil microbes seems to have led to 'The under-recognized dominance of Verrucomicrobia in soil bacterial communities' (Bergmann et al., 2011).
158. Correct answers to factual knowledge questions in physical and biological sciences, by country/region, are reported in *Science and Engineering Indicators*, a report of the US National Science Foundation in 2012 (<http://www.nsf.gov/statistics/seind12/c7/tt07-09.htm>).
159. Craig Venter expressed his concern about the level of public knowledge about genetics in a radio interview with KPBS Radio, 28 October 2013:

> a recent poll has shown that 50% of people don't realize that tomatoes have DNA. So it is difficult to start an intelligent conversation when the fundamental biological knowledge is so limited. So, it means our education system has really failed people. So if you don't realize that tomatoes have DNA, talking about synthetic DNA and creating new things could obviously [be] so fearful to people, versus as I describe in the book, every living organism including humans on this planet, we are DNA, software driven machines. (<http://www.kpbs.org/audioclips/20187/>)

160. The novel interpretation of what a 'lingering kiss' might mean comes from the *New Scientist* in January 2013 (<http://www.newscientist.com/article/mg21729014.900-lingering-kiss-dna-persists-in-the-mouth-after-smooch.html#.U7h6H01dVGU>), reporting on a study of the 'Prevalence and persistence of male DNA identified in mixed saliva samples after intense kissing' (Kamodyová et al., 2013).
161. DNA tagging systems are now in use, for example, from companies like ADNAS in the form of their DNAnet 'Intruder Alert' system that uses fluorescently tagged DNA (<http://www.adnas.com/products/dnanet>).
162. This results from Rob Knight's work on tracking the microbes left behind on inanimate objects. The science was translated into a *CSI* episode and the original article can be found here: <http://www.ncbi.nlm.nih.gov/pubmed/?term=20231444>.
163. See the website of Lisa Matisoo-Smith at the University of Otago for many references related to reconstructing Polynesian migrations: <http://anatomy.otago.ac.nz/index.php?option=com_content&task=view&id=522&Itemid=46>. For a review of recent evidence from chickens in particular, see the following paper: 'Investigating the global dispersal of chickens in prehistory using ancient mitochondrial DNA signatures' (Storey et al., 2012).
164. See 'Identification of Polynesian mtDNA haplogroups in remains of Botocudo Amerindians from Brazil' which reports finding mitochondrial sequences characteristic of Polynesians in DNA extracted from ancient skulls from a now extinct population of Native Americans (Gonçalves et al., 2013). While demonstrating the potential of DNA approaches, various scenarios remain plausible for explaining these data, providing tantalizing but still inconclusive evidence about human history in this region.
165. A study by Caroline Roullier and colleagues provides strong genetic support for the prehistoric introduction of sweet potato from the Peru–Ecuador region into Polynesia; see 'Historical collections reveal patterns of diffusion of sweet potato in Oceania obscured by modern plant movements and recombination' (Roullier et al., 2013). In a news story in *Science* ('Clues to prehistoric human exploration found in sweet potato genome'), Roullier pointed out that while 'the genetic analysis alone doesn't prove that premodern Polynesians made contact with South America, it strongly supports the existing archaeological and linguistic evidence pointing to that conclusion': <http://news.sciencemag.org/biology/2013/01/clues-prehistoric-human-exploration-found-sweet-potato-genome>.
166. Paul Hebert first proposed COI as a 'DNA barcoding gene' for animals in 'Biological identifications through DNA barcodes' (Hebert et al., 2003) and by 2007 had built 'BOLD: the Barcode of Life Data System' (Ratnasingham & Hebert, 2007), a database registry of species with valid DNA barcodes and an associated set of tools: <http://www.boldsystems.org/>.

167. The iBOL tagline is ‘making every species count’ and the homepage of the project can be found here: <http://ibol.org/>.
168. See ‘Screening mammal biodiversity using DNA from leeches’ (Schnell et al., 2012).
169. The story ‘In the soup: a dash of biodiversity’, reported by the *New York Times*, covered the use of DNA barcoding to reveal the presence of species listed as endangered by the International Union for the Conservation of Nature: <http://green.blogs.nytimes.com/2012/08/09/in-shark-fin-soup-a-dash-of-biodiversity/DNA>. Barcoding approaches are increasingly important in conservation, for example, as discussed in the study ‘Applying genetic techniques to study remote shark fisheries in northeastern Madagascar’ (Doukakis et al., 2011).
170. See ‘DNA barcoding of parasitic nematodes: is it kosher?’ (Siddall et al., 2012) and a report on the study by the American Museum of Natural History: <http://www.amnh.org/our-research/science-news/2012/dna-bar coding-of-parasitic-worms-is-it-kosher>.
171. Willerlev’s landmark paper describing Pleistocene communities through ‘dirt genomics’ of ancient DNA: ‘Diverse plant and animal genetic records from Holocene and Pleistocene sediments’ (Willerslev et al., 2003).
172. In their study ‘Meta-barcoding of “dirt” DNA from soil reflects vertebrate biodiversity’, Willerslev and colleagues ‘explored the accuracy and sensitivity of “dirt” DNA as an indicator of vertebrate diversity, from soil sampled at safari parks, zoological gardens and farms with known species compositions’ (Andersen et al., 2012).
173. For a perspective on how eDNA and genetic monitoring might help ‘public agencies implement environmental laws’, see Ryan Kelly and colleagues’ article entitled ‘Harnessing DNA to improve environmental management’ (Kelly et al., 2014).
174. The list is found here: <http://www.issg.org/database/species/search.asp?st=100ss>.
175. DNA monitoring of the invasion of US waters and the threat to the Great Lakes from Asian carp is widely covered in the press (for example, the article at <http://www.scientificamerican.com/article/asian-carp-woes/>). This is one website with general information about the Asian carp and its threat: <http://www.nps.gov/miss/naturescience/ascarpover.htm>.
176. For a review of transgenic techniques in insect control, see ‘Insect transgenesis: current applications and future prospects’ (M. J. Fraser, 2012), and for mosquitoes in particular, see ‘Genetic control of mosquitoes’ (Alphey, 2014).
177. A history of the medfly preventative release programme can be found at the California Department of Food and Agriculture, <http://www.cdfa.ca.gov/plant/pdep/prpinfo/pg1.html>.

178. High temperatures can be lethal to some female medflies with particular genotypes, as discovered by researchers at the International Atomic Energy Agency. See 'For males only: temperature-sensitive medflies', published in the June 2000 issue of *Agricultural Research* magazine.
179. See 'Fluorescent sperm marking to improve the fight against the pest insect Ceratitis capitata' (Scolari et al., 2008).
180. Horizontal gene transfer among microbes is now known to be common and Hehemann's study found gut microbes apparently picking up genes from the food that humans consume: 'Transfer of carbohydrate-active enzymes from marine bacteria to Japanese gut microbiota' (Hehemann et al., 2010).
181. For a general story on the concept of our gut microbes gaining genes from our food, see 'Microbiology: genetic pot luck' (Sonnenburg, 2010).
182. The study 'Illicit drugs, a novel group of environmental contaminants' found that 'residues of drugs of abuse have become widespread surface water contaminants in populated areas' (such as the River Thames). The authors express environmental concerns because 'most of these residues still have potent pharmacological activities' and so 'their presence in the aquatic environment may have potential implications for human health and wildlife' (Zuccato et al., 2008).
183. The title of the study by Larsson and colleagues is 'Pyrosequencing of antibiotic-contaminated river sediments reveals high levels of resistance and gene transfer elements' (Kristiansson et al., 2011).
184. In a *Nature* commentary, Mark Woolhouse and Jeremy Farrar called for the creation of 'an intergovernmental panel on antimicrobial resistance' (Woolhouse & Farrar, 2014). World leaders met in Brussels the following month and appeared to take note. The Declaration from the G7 Summit in June 2014 included a commitment 'to develop a Global Action Plan on antimicrobial resistance' (<http://europa.eu/rapid/press-release_MEMO-14-402_en.htm>). Returning from the summit, British Prime Minister David Cameron announced that 'If we fail to act, we are looking at an almost unthinkable scenario where antibiotics no longer work and we are cast back into the dark ages of medicine where treatable infections and injuries will kill once again' (see <http://www.bbc.com/news/health-28098838>).
185. See 'Phylogenetic structure of the prokaryotic domain: the primary kingdoms' (Woese & Fox, 1977).
186. Jo Handelsman and colleagues coined this term in their article on 'Molecular biological access to the chemistry of unknown soil microbes: a new frontier for natural products' (Handelsman et al., 1998).
187. Two papers in 2004 marked the dawn of environmental shotgun sequencing: Jill Banfield and colleagues published the first in a study on 'Community structure and metabolism through reconstruction of microbial genomes from the environment' (Tyson et al., 2004), targeting a simple

biofilm community hundreds of feet underground in an acid mine drainage. In the second, Craig Venter's team reported their 'Environmental genome shotgun sequencing of the Sargasso Sea' (J Venter et al., 2004), targeting a nutrient limited part of the open ocean that was thought to be relatively devoid of life as a pilot study for the Global Ocean Sampling (GOS) Expedition.

188. The results of the GOS Expedition were published in a special issue of the open access journal *PloS Biology*, including the primary data paper showing the uniqueness of each sampling site: 'The *Sorcerer II* Global Ocean Sampling Expedition: northwest Atlantic through eastern tropical Pacific' (Rusch et al., 2007).
189. The International Census of Marine Microbes (ICOMM) (<http://icomm.mbl.edu/>) launched as part of the larger Census of Marine Life (<http://www.coml.org/>), a 10-year study of the oceans. All data from ICOMM can be found here: <http://vamps.mbl.edu/resources/databases.php>.
190. The homepage of the Earth Microbiome Project (<http://www.earthmicrobiome.org/>) contains access to all data generated thus far.
191. May originally asked the question in his paper 'How many species are there on earth?' and discussed factors influencing numbers and estimates (May, 1988).
192. May posed this question in his paper 'Tropical arthropod species, more or less?' (May, 2010).
193. Mora et al. asked 'How many species are there on Earth and in the ocean?' (Mora et al., 2011).
194. The IIES is now at the State University of New York College of Environmental Health and Forestry (SUNY-ESF) (http://www.esf.edu/species/>) and compiles a Top 10 List of New Species each year. They have also published a coffee table book on the 100 most unusual species. Quentin Wheeler writes a column for *The Guardian* called 'New to nature'.
195. The term 'dark' is increasingly being used for unknown taxa and genes. For example, it was used to describe marine T4-type bacteriophages, as 'a ubiquitous component of the dark matter of the biosphere' (Filée et al., 2005). More recently, Rod Page used it to lament the lack of species names for DNA sequences in GenBank, even for mammals. In his blog post 'Dark taxa: GenBank in a post-taxonomic world', Page considers 'a post-taxonomic world where taxonomic names ... are not that important'. He points out that microbiology seems to be doing fine as a discipline even if 'In 2010 less than 1% of newly sequenced bacteria had been formerly described' (see <http://iphylo.blogspot.com/2011/04/dark-taxa-genbank-in-post-taxonomic.html>).
196. The origin of so-called orphan genes, those that are not found in other species or lineages, remains one of the great mysteries of genomics. One study on the 'Origin of primate orphan genes: a comparative genomics

approach' found that 'around 53% of the orphan genes contain sequences derived from transposable elements' (Toll-Riera et al., 2009).

197. Asking whether we can 'name Earth's species before they go extinct', May and colleagues concluded that 'with modestly increased effort in taxonomy and conservation, most species could be discovered and protected from extinction' (M. J. Costello, May, & Stork, 2013). Mora et al. responded, however, by claiming that this was 'overly optimistic because of a limited selection and interpretation of available evidence that tends to overestimate rates of species description and underestimate the number of species on Earth and their current extinction rate' (Mora, Rollo, & Tittensor, 2013). Stuart Pimm and colleagues estimated that current extinction rates of eukaryotic species are at least 1,000 times the background rate in a 2014 review of biodiversity status,, and that is probably still an underestimate (Pimm et al., 2014).
198. Barnosky et al. asked: 'Has the Earth's sixth mass extinction already arrived?' They concluded that 'the recent loss of species is dramatic and serious but does not yet qualify as a mass extinction'. However, they cautioned that 'additional losses of species in the 'endangered' and 'vulnerable' categories could accomplish the sixth mass extinction in just a few centuries' (Barnosky et al., 2011).
199. For an example, see 'Science and government. Navigating the anthropocene: improving Earth system governance' (Biermann et al., 2012).
200. The book *Abundance: The Future Is Better Than You Think* (Diamandis, 2012) takes a very positive view of now and the future, thanks to accelerating advances in science and technology: <http://www.abundancethebook.com/>.
201. The book *Ten Billion* takes a dark view of the future and humanity's capacity to successfully navigate the consequences of its growing population and demands on the environment (Emmott, 2013).
202. UC Davis professor Jonathan Eisen, a leader in the field, has reviewed the use of the suffix 'omics' that has exploded since the Human Genome Project. Recognizing the power of the underlying approach, he points out that merely 'adding "ome" or "omics" onto some term does not suddenly make it "genomic-y"'. He criticizes the overuse of the suffix as what he calls 'badomics' (Eisen, 2012) and calls out offenders on his blog: <http://phylogenomics.blogspot.com/search/label/bad%20omics%20word%20of%20the%20day>.
203. In a 2014 article entitled 'Individualized medicine from prewomb to tomb', Eric Topol provides a comprehensive review of the state of the art of personalized (or as he prefers 'individualized') medicine (Topol, 2014), including a discussion of the 'synderome' (Chen et al., 2012).
204. A later article in the *New York Times* on 'personal data projects' was entitled 'The Data Driven Life': <http://www.nytimes.com/2010/05/02/magazine/02self-measurement-t.html?pagewanted=all&_r=0>.

205. Larry Smarr's effort to study his microbiome has been widely covered in the media. 'Quantifying your body: a how-to guide from a systems biology perspective' (Smarr, 2012) describes the start of his self-study and the benefits of quantified-self approaches.
206. Leroy Hood is the president of the Institute for Systems Biology (ISB), which has founded the P4 Institute (<http://p4mi.org/>), and has written widely about his P4 vision for the future.
207. See 'Host lifestyle affects human microbiota on daily timescales' (David et al., 2014).
208. See 'Significant changes in the skin microbiome mediated by the sport of roller derby' (Meadow et al., 2013).
209. The Sloan Foundation is supporting basic research on the indoor microbiome through its 'Microbiology of the Built Environment' programme, see <http://www.sloan.org/major-program-areas/basic-research/mobe/?L=0%3Ftx_solr%5Bpage%5D%3D1>.
210. The Hospital Microbiome Project aims to 'characterize the taxonomic composition of surface-, air-, water-, and human-associated microbial communities in two hospitals to monitor changes in community structure following the introduction of patients and hospital staff'. See the project website: <http://hospitalmicrobiome.com/>.
211. The Hospital Microbiome team documented a first pass through the construction site for the University of Chicago's New Hospital Pavilion. What they found was discussed on their website: <http://hospitalmicrobiome.com/construction-samples/>.
212. See 'The L4 time-series: the first 20 years' (Harris, 2010).
213. The results of the six-year time-series study of the Western Channel Observatory L4 buoy off Plymouth were presented in a paper, 'Defining seasonal marine microbial community dynamics' (Gilbert et al., 2012), and expanded with deep sequencing of some of the samples that presented 'Evidence for a persistent microbial seed bank throughout the global ocean' (Gibbons et al., 2013).
214. The statement 'Everything is everywhere: but the environment selects' on the 'ubiquitous distribution and ecological determinism in microbial biogeography' is credited to Dutch microbiologist 'Martinus Wilhelm Beijerinck early in the twentieth century and specifically articulated in 1934 by his compatriot, Lourens G. M. Baas Becking' (O'Malley, 2008).
215. According to the team carrying out the study, the black cottonwood *Populus trichocarpa* was 'selected as the model forest species for genome sequencing not only because of its modest genome size but also because of its rapid growth, relative ease of experimental manipulation, and range of available genetic tools' (Tuskan et al., 2006).
216. See 'Extending genomics to natural communities and ecosystems' (Whitham et al., 2008).

217. See 'New frontiers in community and ecosystem genetics for theory, conservation, and management' (Bailey & Genung, 2012).
218. See *The Extended Phenotype: The Long Reach of the Gene* (Dawkins, 1982).
219. See 'Draft genome sequence of the Coccolithovirus Emiliania huxleyi virus 203' (Nissimov et al., 2011) and 'Pan genome of the phytoplankton Emiliania underpins its global distribution' (Read et al., 2013).
220. The home page of the cod genome project describes its selection for commercial domestication <http://www.codgenome.no/> and the primary genome report was published in *Nature*: 'The genome sequence of Atlantic cod reveals a unique immune system' (Star et al., 2011).
221. The first coral genome to be sequenced—'Using the Acropora digitifera genome to understand coral responses to environmental change'—provides a platform for understanding the molecular basis of symbiosis and responses to environmental changes (Shinzato et al., 2011).
222. 'Draft assembly of the Symbiodinium minutum nuclear genome reveals dinoflagellate gene structure' provides the first genomic view of this important group of coral endosymbionts (Shoguchi et al., 2013).
223. The Scripps Institution of Oceanography maintains an informative website dedicated to the Keeling Curve and a daily record of atmospheric carbon dioxide concentrations, see <http://keelingcurve.ucsd.edu/>.
224. Whitham and colleagues have reported findings that suggest 'emergent ecological properties such as stability can be genetically based and thus subject to natural selection' (Keith, Bailey, & Whitham, 2010). For a detailed description of the case of the pinyon pine, see 'Deadly combination of genes and drought: increased mortality of herbivore-resistant trees in a foundation species' (Sthultz, Gehring, & Whitham, 2009).
225. See 'Genomic basis for coral resilience to climate change' (Barshis et al., 2013).
226. See 'Climate-change adaptation: designer reefs' (Mascarelli, 2014).
227. *A World in One Cubic Foot: Portraits of Biodiversity* (Liittschwager, 2012).
228. A news story 'Treasure island: pinning down a model ecosystem' in *Nature* described the rationale behind the Moorea Biocode Project with 'every species on paradise isle to be catalogued' (Check, 2006).
229. Craig Venter sampled marine microbes in the waters off Moorea in his Global Ocean Survey and Linda Amaral Zettler of the Marine Biological Laboratory in Woods Hole conducted the first microbial inventory of a coral reef across the domains of life on Moorea as part of her LTER MIRADAS project (McCliment et al., 2011).
230. The broad impacts of big data and 'datafication'—not least on science—are described by Kenneth Cukier and Viktor Mayer-Schonberger in their thought-provoking book *Big Data: A Revolution That Will Transform How We Live, Work, and Think* (Cukier & Mayer-Schönberger, 2013).
231. See the Moorea Island Digital Ecosystem Avatar (IDEA) Project homepage: <http://mooreaIDEA.org>.

232. In this paper, Drew Purves of Microsoft Research and colleagues outlined their vision for GEMs (Purves et al., 2013).
233. The IPBES is 'an independent intergovernmental body open to all member countries of the United Nations. The members are committed to building IPBES as the leading intergovernmental body for assessing the state of the planet's biodiversity, its ecosystems and the essential services they provide to society'. See the IPBES homepage for details: <http://www.ipbes.net/>.
234. A call for a 'Genomic Observatories Network' to monitor the Earth was first published in *Nature* in 2012 (Davies, Field, & Network, 2012). The founding charter of the network laying out its mission and initial members was published just over a year later (Davies et al., 2014) following a series of international meetings under the auspices of the Genomic Standards Consortium (GSC) and the Group on Earth Observations (GEO) Biodiversity Observation Network (GEO-BON).
235. See 'A whole-cell computational model predicts phenotype from genotype' (Karr et al., 2012). The *New York Times* (20 July 2012) reported the breakthrough as follows: 'Scientists at Stanford University and the J. Craig Venter Institute have developed the first software simulation of an entire organism, a humble single-cell bacterium that lives in the human genital and respiratory tracts' (<http://www.nytimes.com/2012/07/21/science/in-a-first-an-entire-organism-is-simulated-by-software.html?_r=2&hp&>).
236. In 2013, replacing problematic mitochondria with those of a healthy donor was moving towards regulatory approval in the UK and the USA; see the *New York Times* article 'Three Biological Parents and a Baby' (<http://well.blogs.nytimes.com/2013/12/16/three-biological-parents-and-a-baby/?_php=true&_type=blogs&_r=0>).
237. The concept of genetic effects on communities and ecosystems has been explored theoretically and empirically by Tom Whitham and colleagues, exploiting long-standing common garden approaches in forestry and combining them with genomics (Shuster et al., 2006; Whitham et al., 2008).
238. Dawkins used the more general term 'replicator' instead of gene, perhaps implying an ambition to extend the principle to 'memes' and cultural evolution. In that event, this might be a basis for a 'scientific theory of human progress'. It could explain why society has tended towards greater cooperation and more respect for individual human rights.
239. In his classic paper, Dobzhansky describes how evolution and genetics make sense of life: 'The unity of life is no less remarkable than its diversity. Most forms of life are similar in many respects. The universal biologic similarities are particularly striking in the biochemical dimension. From viruses to man, heredity is coded in just two, chemically related substances: DNA and RNA. The genetic code is as simple as it is universal. There are only four genetic "letters" in DNA: adenine, guanine, thymine, and cytosine. Uracil replaces thymine in RNA. The entire evolutionary development

of the living world has taken place not by invention of new "letters" in the genetic "alphabet" but by elaboration of ever-new combinations of these letters.' (Dobzhansky, 1973).

240. In their study of 'Genomic diversity and evolution of the head crest in the rock pigeon', Jun Shang of the BGI and colleagues sequenced 38 *Columba livia* individuals including 36 domestic breeds and 2 feral birds (Shapiro et al., 2013). Their subsequent analyses confirmed Darwin's hypothesis that this species gave rise to the ornamental breeds he loved so much; see 'Pigeon DNA proves Darwin right' (Humphries, 2013).

241. The Sunjammer website describes the mission as 'NASA's first solar sail mission to deep space and the largest sail ever flown. As a NASA Technology Demonstration Mission, Sunjammer is the final step before solar sails are integrated into future space missions. Onboard Sunjammer, advanced technologies will provide Earth with its earliest warning to date of potentially hazardous solar activity. Additionally, the Cosmic Archive will carry a message of hope for future generations.' See <http://www.sunjammermission.com/>. *The Guardian* published a story entitled 'Arthur C Clarke's DNA to join mission into deep space' in June 2013. See <http://www.theguardian.com/books/2013/jun/26/arthur-c-clarke-hair-deep-space>.

242. Nick Loman's blog is at <http://pathogenomics.bham.ac.uk/blog/2013/12/the-biggest-genome-sequencing-projects-the-uber-list/>.

243. See 'Evolutionary pathway to increased virulence and epidemic group A Streptococcus disease derived from 3,615 genome sequences' (Nasser et al., 2014).

244. See <http://www.cnet.com/news/ibm-sees-big-opportunity-in-sequencing-microbes/>.

245. See <http://www.biodiversity.uoguelph.ca/>.

246. On its website, the X Prize Foundation states that it 'developed the Qualcomm Tricorder XPRIZE to spur radical innovation in personal health care technology. The competition is designed to address the inefficient, expensive, and inertia-bound healthcare system in the United States and elsewhere' (<http://www.qualcommtricorderxprize.org/>).

247. See Paul Edwards's excellent account of the global study of weather and climate in his book *The Vast Machine* (Edwards, 2010).

248. In the study 'Pattern and synchrony of gene expression among sympatric marine microbial populations', an *in situ* robotic sampler was used to continuously track and repeatedly sample microbial populations as they drifted through the water column. The samples were then subject to genome-wide transcriptome profiling through RNA sequencing to investigate gene expression patterns through the community (Ottesen et al., 2013). Chris Scholin has also written an accessible review of the development of what he champions as 'ecogenomic sensors' (Scholin, 2013).

249. See the International Barcode of Life website for link to DNA barcoding projects: <http://ibol.org/>.
250. See the article in *Nature* about Tara Oceans: 'Systems ecology: biology on the high seas' (Ainsworth, 2013).
251. The Genographic homepage describes the rationale, history, and progress to date of this revolutionary project and offers self-study kits for purchase: <http://genographic.nationalgeographic.com/>.
252. See 'An African American paternal lineage adds an extremely ancient root to the human Y chromosome phylogenetic tree' (Mendez et al., 2013).
253. The original uBiome crowd-funding campaign can be seen on Indiegogo website (<http://www.indiegogo.com/projects/ubiome-sequencing-your-microbiome>) and the company now sells kits off their main company website: <http://ubiome.com/>.
254. The American Gut project is on the Human Food Project webpage: <http://humanfoodproject.com/americangut/>. The crowd-funding campaign website is <http://www.indiegogo.com/projects/american-gut-what-s-in-your-gut--7>.
255. The book describes the human relationship with food, history, and our microbiome (Leach, 2012).
256. Rob Dunn—author of *The Wild Life of Our Bodies* (Dunn, 2011)—and colleagues published a study of bellybuttons: 'A jungle in there: bacteria in belly buttons are highly diverse, but predictable' (Hulcr et al., 2012). We have a whole world in our navel.
257. Writing for *Discover Magazine*, Carl Zimmer happily reported his own contribution to Dunn's study; see 'On the Occasion of My Belly Button Entering the Scientific Literature' <http://blogs.discovermagazine.com/loom/2012/11/07/on-the-occasion-of-my-belly-button-entering-the-scientific-literature/>.
258. See <http://homes.yourwildlife.org/>.
259. Darwin landed in Tahiti in 1834. He stepped ashore at Matavai Bay, onto the same volcanic black sand Captain James Cook trod in 1769. Climbing the mountains of Tahiti, Darwin looked to the north-west and was struck by the sight of Moorea about 10 miles away. He noted in his journal that 'the effect was very pleasing' and he compared the island to a picture, with the island as the drawing, the lagoon the marginal paper, and the breakers on the reef the frame (Darwin, 1839). Such observations helped Darwin develop an evolutionary hypothesis for how geological and biological processes combined to shape 'coral-islands'. In 1842, he published *The Structure and Distribution of Coral Reefs*, setting out this theory of barrier-reef and atoll formation (Darwin, 1842), which is still largely held as correct today.
260. See Andrea Wulf's book on the Transit of Venus as the first globally coordinated scientific study: *Chasing Venus: The Race to Measure the Heavens* (Wulf, 2013).

REFERENCES

Abecasis, G. R., Auton, A., Brooks, L. D., DePristo, M. A., Durbin, R. M., Handsaker, R. E., ... McVean, G. A. (2012). An integrated map of genetic variation from 1,092 human genomes. *Nature*, 491(7422), 56–65. doi:10.1038/nature11632

Ainsworth, C. (2013). Systems ecology: biology on the high seas. *Nature*, 501(7465), 20–3. doi:10.1038/501020a

Allentoft, M. E., Collins, M., Harker, D., Haile, J., Oskam, C. L., Hale, M. L., ... Bunce, M. (2012). The half-life of DNA in bone: measuring decay kinetics in 158 dated fossils. *Proceedings. Biological Sciences/The Royal Society*, 279(1748), 4724–33. doi:10.1098/rspb.2012.1745

Alphey, L. (2014). Genetic control of mosquitoes. *Annual Review of Entomology*, 59(October), 205–24. doi:10.1146/annurev-ento-011613-162002

Altschul, S., Gish, W., Miller, W., Meyers, E., & Lippman, D. (1990). Basic local alignment search tool. *Journal of Molecular Biology*, 215(3), 403–10. doi:http://dx.doi.org/10.1016/S0022-2836(05)80360-2

Amborella, G. P. (2013). The Amborella genome and the evolution of flowering plants. *Science* (New York, NY), 342(6165), 1241089. doi:10.1126/science.1241089

Andersen, K., Bird, K. L., Rasmussen, M., Haile, J., Breuning-Madsen, H., Kjaer, K. H., ... Willerslev, E. (2012). Meta-barcoding of 'dirt' DNA from soil reflects vertebrate biodiversity. *Molecular Ecology*, 21(8), 1966–79. doi:10.1111/j.1365-294X.2011.05261.x

Anderson, S., Bankier, A. T., Barrell, B. G., de Bruijn, M. H. L., Coulson, A. R., Drouin, J., ... Young, I. G. (1981). Sequence and organization of the human mitochondrial genome. *Nature*, 290(5806), 457–65. doi:10.1038/290457a0

Anker, S., & Nelkin, D. (2003). *The Molecular Gaze: Art in the Genetic Age*. Cold Spring Harbor, NY: Cold Spring Harbor Laboratory Press.

Annas, G. J., & Elias, S. (2014). 23andMe and the FDA. *New England Journal of Medicine*, 370(11), 985–8. doi:10.1056/NEJMp1316367

Aparicio, S., Chapman, J., Stupka, E., Putnam, N., Chia, J.-M., Dehal, P., ... Brenner, S. (2002). Whole-genome shotgun assembly and analysis of the genome of Fugu rubripes. *Science* (New York, NY), 297(5585), 1301–10. doi:10.1126/science.1072104

Arabidopsis Genome Initiative (2000). Analysis of the genome sequence of the flowering plant Arabidopsis thaliana. *Nature*, 408(6814), 796–815. doi:10.1038/35048692

Bailey, J., & Genung, M. (2012). New frontiers in community and ecosystem genetics for theory, conservation, and management. *New Phytologist*, 193(1), 24–6. Retrieved from <http://onlinelibrary.wiley.com/doi/10.1111/j.1469-8137.2011.03973.x/full>.

Barnosky, A. D., Matzke, N., Tomiya, S., Wogan, G. O. U., Swartz, B., Quental, T. B.,...Ferrer, E. A. (2011). Has the Earth's sixth mass extinction already arrived? *Nature*, 471(7336), 51–7. doi:10.1038/nature09678

Barshis, D. J., Ladner, J. T., Oliver, T. A., Seneca, F. O., Traylor-Knowles, N., & Palumbi, S. R. (2013). Genomic basis for coral resilience to climate change. *Proceedings of the National Academy of Sciences of the United States of America*, 110(4), 1387–92. doi:10.1073/pnas.1210224110

Bennett, G. M., & Moran, N. A. (2013). Small, smaller, smallest: the origins and evolution of ancient dual symbioses in a Phloem-feeding insect. *Genome Biology and Evolution*, 5(9), 1675–88. doi:10.1093/gbe/evt118

Bergmann, G. T., Bates, S. T., Eilers, K. G., Lauber, C. L., Caporaso, J. G., Walters, W. A.,...Fierer, N. (2011). The under-recognized dominance of Verrucomicrobia in soil bacterial communities. *Soil Biology & Biochemistry*, 43(7), 1450–5. doi:10.1016/j.soilbio.2011.03.012

Bernstein, B. E., Birney, E., Dunham, I., Green, E. D., Gunter, C., & Snyder, M. (2012). An integrated encyclopedia of DNA elements in the human genome. *Nature*, 489(7414), 57–74. doi:10.1038/nature11247

Biermann, F., Abbott, K., Andresen, S., Bäckstrand, K., Bernstein, S., Betsill, M. M., ...Zondervan, R. (2012). Science and government. Navigating the anthropocene: improving Earth system governance. *Science* (New York, NY), 335(6074), 1306–7. doi:10.1126/science.1217255

Blaser, M. J. (2014). *Missing Microbes: How the Overuse of Antibiotics Is Fueling Our Modern Plagues*. New York: Henry Holt and Co.

Bult, C. J., White, O., Olsen, G. J., Zhou, L., Fleischmann, R. D., Sutton, G. G., ... Venter, J. C. (1996). Complete genome sequence of the methanogenic archaeon, *Methanococcus jannaschii*. *Science*, 273(5278), 1058–73. doi:10.1126/science.273.5278.1058

C. elegans Sequencing Consortium (1998). Genome sequence of the nematode C. elegans: a platform for investigating biology. *Science*, 282(5396), 2012–18. doi:10.1126/science.282.5396.2012

Carey, N. (2013). *The Epigenetics Revolution: How Modern Biology Is Rewriting Our Understanding of Genetics, Disease, and Inheritance*. New York: Columbia University Press.

Carlton, J. M., Hirt, R. P., Silva, J. C., Delcher, A. L., Schatz, M., Zhao, Q., ...Johnson, P. J. (2007). Draft genome sequence of the sexually transmitted pathogen *Trichomonas vaginalis*. *Science* (New York, NY), 315(5809), 207–12. doi:10.1126/science.1132894

Chan, W. F. N., Gurnot, C., Montine, T. J., Sonnen, J. A., Guthrie, K. A., & Nelson, J. L. (2012). Male microchimerism in the human female brain. *PloS One*, 7(9), e45592. doi:10.1371/journal.pone.0045592

Check, E. (2006). Treasure island: pinning down a model ecosystem. *Nature*, 439(7075), 378–9. doi:10.1038/439378a

Chen, R., Mias, G. I., Li-Pook-Than, J., Jiang, L., Lam, H. Y. K., Chen, R., ... Snyder, M. (2012). Personal omics profiling reveals dynamic molecular and medical phenotypes. *Cell*, 148(6), 1293–307. doi:10.1016/j.cell.2012.02.009

Church, G., & Regis, E. (2012). *Regenesis: How Synthetic Biology Will Reinvent Nature and Ourselves*. New York: Basic Books.

Church, G. M. (2005). The Personal Genome Project. *Molecular Systems Biology*, 1, 2005.0030. doi:10.1038/msb4100040

Ciccarelli, F. D., Doerks, T., von Mering, C., Creevey, C. J., Snel, B., & Bork, P. (2006). Toward automatic reconstruction of a highly resolved tree of life. *Science* (New York, NY), 311(5765), 1283–7. doi:10.1126/science.1123061

Claes, P., Liberton, D. K., Daniels, K., Rosana, K. M., Quillen, E. E., Pearson, L. N., ... Shriver, M. D. (2014). Modeling 3D facial shape from DNA. *PLoS Genetics*, 10(3), e1004224. doi:10.1371/journal.pgen.1004224

Cochrane, G., Cook, C. E., & Birney, E. (2012). The future of DNA sequence archiving. *GigaScience*, 1(1), 2. doi:10.1186/2047-217X-1-2

Costello, E. K., Stagaman, K., Dethlefsen, L., Bohannan, B. J. M., & Relman, D. A. (2012). The application of ecological theory toward an understanding of the human microbiome. *Science* (New York, NY), 336(6086), 1255–62. doi:10.1126/science.1224203

Costello, M. J., May, R. M., & Stork, N. E. (2013). Can we name Earth's species before they go extinct? *Science* (New York, NY), 339(6118), 413–16. doi:10.1126/science.1230318

Cukier, K., & Mayer-Schönberger, V. (2013). *Big Data: A Revolution That Will Transform How We Live, Work, and Think*. New York: Eamon Dolan/Houghton Mifflin Harcourt.

Darwin, C. (1839). *Journal of Researches into the Natural History and Geology of the Countries Visited During the Voyage of H.M.S. Beagle Round the World. Narrative of the Surveying Voyages of His Majesty's Ships Adventure and Beagle between the Years 1826 and 1836, Describing Their Examination of the Southern Shores of South America, and the Beagle's Circumnavigation of the Globe. Journal and Remarks. 1832*. London: Henry Colburn. Retrieved from <http://darwin-online.org.uk/>.

Darwin, C. (1842). *The Structure and Distribution of Coral Reefs. Being the first part of the geology of the voyage of the Beagle, under the command of Capt. Fitzroy, R.N. during the years 1832 to 1836*. London: Smith Elder and Co.

David, L. A., Materna, A. C., Friedman, J., Campos-Baptista, M. I., Blackburn, M. C., Perrotta, A., ... Alm, E. J. (2014). Host lifestyle affects human microbiota on daily timescales. *Genome Biology*, 15(7), R89. doi:10.1186/gb-2014-15-7-r89

Davies, N., Field, D., Amaral-Zettler, L., Clark, M. S., Deck, J., Drummond, A., ... Zingone, A. (2014). The founding charter of the Genomic Observatories Network. *GigaScience*, 3(1), 2. doi:10.1186/2047-217X-3-2

Davies, N., Field, D., & Genomic Observatories Network (2012). A genomic network to monitor Earth. *Nature*, 481(7380), 145. doi:10.1038/481145a

Dawkins, R. (1976). *The Selfish Gene*. New York: Oxford University Press.

Dawkins, R. (1982). *The Extended Phenotype: The Long Reach of the Gene*. Oxford: Oxford University Press.

Del Campo, J., Sieracki, M. E., Molestina, R., Keeling, P., Massana, R., & Ruiz-Trillo, I. (2014). The others: our biased perspective of eukaryotic genomes. *Trends in Ecology & Evolution*, 29(5), 252–9. doi:10.1016/j.tree.2014.03.006

Diamandis, P. H. (2012). *Abundance: The Future Is Better Than You Think*. New York: Free Press.

Dobzhansky, T. (1973). Nothing in biology makes sense except in the light of evolution. *American Biology Teacher*, 35(3), 125–9. doi:10.2307/4444260

Domazet-Loso, T., & Tautz, D. (2008). An ancient evolutionary origin of genes associated with human genetic diseases. *Molecular Biology and Evolution*, 25(12), 2699–707. doi:10.1093/molbev/msn214

Doukakis, P., Hanner, R., Shivji, M., Bartholomew, C., Chapman, D., Wong, E., & Amato, G. (2011). Applying genetic techniques to study remote shark fisheries in northeastern Madagascar. *Mitochondrial DNA*, 22 Suppl 1(October), 15–20. doi:10.3109/19401736.2010.526112

Draus-Barini, J., Walsh, S., Pośpiech, E., Kupiec, T., Głąb, H., Branicki, W., & Kayser, M. (2013). Bona fide colour: DNA prediction of human eye and hair colour from ancient and contemporary skeletal remains. *Investigative Genetics*, 4(1), 3. doi:10.1186/2041-2223-4-3

Dunn, R. (2011). *The Wild Life of Our Bodies*. New York: Harper.

Eckburg, P. B., Bik, E. M., Bernstein, C. N., Purdom, E., Dethlefsen, L., Sargent, M., ... Relman, D. A. (2005). Diversity of the human intestinal microbial flora. *Science* (New York, NY), 308(5728), 1635–8. doi:10.1126/science.1110591

Edwards, P. N. (2010). *The Vast Machine: Computer Models, Climate Data, and the Politics of Global Warming*. Cambridge, MA: MIT Press.

Eisen, J. A. (2012). Badomics words and the power and peril of the ome-meme. *GigaScience*, 1(1), 6. doi:10.1186/2047-217X-1-6

Emmott, S. (2013). *Ten Billion*. New York: Vintage.

Fierer, N., Ladau, J., Clemente, J. C., Leff, J. W., Owens, S. M., Pollard, K. S., ... McCulley, R. L. (2013). Reconstructing the microbial diversity and function of pre-agricultural tallgrass prairie soils in the United States. *Science*, 342(6158), 621–4. doi:10.1126/science.1243768

Filée, J., Tétart, F., Suttle, C. A., & Krisch, H. M. (2005). Marine T4-type bacteriophages, a ubiquitous component of the dark matter of the biosphere. *Proceedings of the National Academy of Sciences of the United States of America*, 102(35), 12471–6. doi:10.1073/pnas.0503404102

Fleischmann, R., Adams, M., White, O., Clayton, R., Kirkness, E., Kerlavage, A., et al. (1995). Whole-genome random sequencing and assembly of *Haemophilus influenzae* Rd. *Science*, 269(5223), 496–512. doi:10.1126/science.7542800

Fraser, C. M., Gocayne, J. D., White, O., Adams, M. D., Clayton, R. A., Fleischmann, R. D., . . . Lucier, T. S. (1995). The minimal gene complement of Mycoplasma genitalium. *Science*, 270(5235), 397–404. doi:10.1126/science.270.5235.397

Fraser, M. J. (2012). Insect transgenesis: current applications and future prospects. *Annual Review of Entomology*, 57, 267–89. doi:10.1146/annurev.ento.54.110807.090545

G10K Community of Scientists (2009). Genome 10K: a proposal to obtain whole-genome sequence for 10 000 vertebrate species. *Journal of Heredity*, 100(6), 659–74. doi:10.1093/jhered/esp086

Gibbons, S. M., Caporaso, J. G., Pirrung, M., Field, D., Knight, R., & Gilbert, J. A. (2013). Evidence for a persistent microbial seed bank throughout the global ocean. *Proceedings of the National Academy of Sciences of the United States of America*, 110(12), 4651–5. doi:10.1073/pnas.1217767110

Gibson, D. G., Benders, G. A, Andrews-Pfannkoch, C., Denisova, E. A, Baden-Tillson, H., Zaveri, J., . . . Smith, H. O. (2008). Complete chemical synthesis, assembly, and cloning of a Mycoplasma genitalium genome. *Science* (New York, NY), 319(5867), 1215–20. doi:10.1126/science.1151721

Gibson, D. G., Glass, J. I., Lartigue, C., Noskov, V. N., Chuang, R.-Y., Algire, M. A., . . . Venter, J. C. (2010). Creation of a bacterial cell controlled by a chemically synthesized genome. *Science* (New York, NY), 329(5987), 52–6. doi:10.1126/science.1190719

Gilbert, J. A., Steele, J. A., Caporaso, J. G., Steinbrück, L., Reeder, J., Temperton, B., . . . Field, D. (2012). Defining seasonal marine microbial community dynamics. *ISME Journal*, 6(2), 298–308. doi:10.1038/ismej.2011.107

Gill, S. R., Pop, M., Deboy, R. T., Eckburg, P. B., Turnbaugh, P. J., Samuel, B. S., . . . Nelson, K. E. (2006). Metagenomic analysis of the human distal gut microbiome. *Science* (New York, NY), 312(5778), 1355–9. doi:10.1126/science.1124234

Goffeau, A., Barrell, B. G., Bussey, H., Davis, R. W., Dujon, B., Feldmann, H., . . . Oliver, S. G. (1996). Life with 6000 genes. *Science*, 274(5287), 546–67. doi:10.1126/science.274.5287.546

Goldman, N., Bertone, P., Chen, S., Dessimoz, C., LeProust, E. M., Sipos, B., & Birney, E. (2013). Towards practical, high-capacity, low-maintenance information storage in synthesized DNA. *Nature*, 494(7435), 77–80. doi:10.1038/nature11875

Gonçalves, V. F., Stenderup, J., Rodrigues-Carvalho, C., Silva, H. P., Gonçalves-Dornelas, H., Líryo, A., . . . Pena, S. D. J. (2013). Identification of Polynesian mtDNA haplogroups in remains of Botocudo Amerindians from Brazil. *Proceedings of the National Academy of Sciences of the United States of America*, 110(16), 6465–9. doi:10.1073/pnas.1217905110

Gullans, S., & Enriquez, J. (2011). *Homo Evolutis*. New York: TED Books.

Haase, C. M., Saslow, L. R., Bloch, L., Saturn, S. R., Casey, J. J., Seider, B. H., ... Levenson, R. W. (2013). The 5-HTTLPR polymorphism in the serotonin transporter gene moderates the association between emotional behavior and changes in marital satisfaction over time. *Emotion* (Washington, DC), 13(6), 1068–79. doi:10.1037/a0033761

Handelsman, J., Rondon, M. R., Brady, S. F., Clardy, J., & Goodman, R. M. (1998). Molecular biological access to the chemistry of unknown soil microbes: a new frontier for natural products. *Chemistry & Biology*, 5(10), R245–9. Retrieved from <http://www.ncbi.nlm.nih.gov/pubmed/9818143>.

Harris, R. (2010). The L4 time-series: the first 20 years. *Journal of Plankton Research*, 32(5), 577–83. doi:10.1093/plankt/fbq021

Hebert, P. D. N., Cywinska, A., Ball, S. L., & deWaard, J. R. (2003). Biological identifications through DNA barcodes. *Proceedings. Biological Sciences/The Royal Society*, 270(1512), 313–21. doi:10.1098/rspb.2002.2218

Hehemann, J.-H., Correc, G., Barbeyron, T., Helbert, W., Czjzek, M., & Michel, G. (2010). Transfer of carbohydrate-active enzymes from marine bacteria to Japanese gut microbiota. *Nature*, 464(7290), 908–12. doi:10.1038/nature08937

Hilbert, M., & López, P. (2011). The world's technological capacity to store, communicate, and compute information. *Science* (New York, NY), 332(6025), 60–5. doi:10.1126/science.1200970

Hillier, L. W., Coulson, A., Murray, J. I., Bao, Z., Sulston, J. E., & Waterston, R. H. (2005). Genomics in *C. elegans*: so many genes, such a little worm. *Genome Research*, 15(12), 1651–60. doi:10.1101/gr.3729105

Hulcr, J., Latimer, A. M., Henley, J. B., Rountree, N. R., Fierer, N., Lucky, A., ... Dunn, R. R. (2012). A jungle in there: bacteria in belly buttons are highly diverse, but predictable. *PloS One*, 7(11), e47712. doi:10.1371/journal.pone.0047712

Humphries, C. (2013). Pigeon DNA proves Darwin right. *Nature News* (31 Jan.). doi:10.1038/nature.2013.12334

Hutchison III, C. A. (1999). Global transposon mutagenesis and a minimal Mycoplasma genome. *Science*, 286(5447), 2165–9. doi:10.1126/science.286.5447.2165

Jiang, J., Jing, Y., Cost, G. J., Chiang, J.-C., Kolpa, H. J., Cotton, A. M., ... Lawrence, J. B. (2013). Translating dosage compensation to trisomy 21. *Nature*, 500(7462), 296–300. doi:10.1038/nature12394

Kaiser, J. (2013). Genomics. Researchers to explore promise, risks of sequencing newborns' DNA. *Science* (New York, NY), 341(6151), 1163. doi:10.1126/science.341.6151.1163

Kalf, R. R. J., Mihaescu, R., Kundu, S., de Knijff, P., Green, R. C., & Janssens, A. C. J. W. (2014). Variations in predicted risks in personal genome testing for common complex diseases. *Genetics in Medicine: Official Journal of the American College of Medical Genetics*, 16(1), 85–91. doi:10.1038/gim.2013.80

Kamodyová, N., Durdiaková, J., Celec, P., Sedláčková, T., Repiská, G., Sviežená, B., & Minárik, G. (2013). Prevalence and persistence of male DNA identified in mixed saliva samples after intense kissing. *Forensic Science International. Genetics*, 7(1), 124–8. doi:10.1016/j.fsigen.2012.07.007

Karr, J. R., Sanghvi, J. C., Macklin, D. N., Gutschow, M. V, Jacobs, J. M., Bolival, B., ... Covert, M. W. (2012). A whole-cell computational model predicts phenotype from genotype. *Cell*, 150(2), 389–401. doi:10.1016/j.cell.2012.05.044

Keith, A. R., Bailey, J. K., & Whitham, T. G. (2010). A genetic basis to community repeatability and stability. *Ecology*, 91(11), 3398–406. Retrieved from <http://www.ncbi.nlm.nih.gov/pubmed/21141200>.

Kelly, R. P., Port, J. A., Yamahara, K. M., Martone, R. G., Lowell, N., Thomsen, P. F., ... Crowder, L. B. (2014). Harnessing DNA to improve environmental management. *Science*, 344(6191), 1455–6. doi:10.1126/science.1251156

Khurana, E., Fu, Y., Colonna, V., Mu, X. J., Kang, H. M., Lappalainen, T., ... Gerstein, M. (2013). Integrative annotation of variants from 1092 humans: application to cancer genomics. *Science* (New York, NY), 342(6154), 1235587. doi:10.1126/science.1235587

King, N., Westbrook, M. J., Young, S. L., Kuo, A., Abedin, M., Chapman, J., ... Rokhsar, D. (2008). The genome of the choanoflagellate *Monosiga brevicollis* and the origin of metazoans. *Nature*, 451(7180), 783–8. doi:10.1038/nature06617

Knight, R., & Buhler, B. (2015). *Follow Your Gut: The Enormous Impact of Tiny Microbes*. New York: Simon & Schuster/TED.

Kristiansson, E., Fick, J., Janzon, A., Grabic, R., Rutgersson, C., Weijdegård, B., ... Larsson, D. G. J. (2011). Pyrosequencing of antibiotic-contaminated river sediments reveals high levels of resistance and gene transfer elements. *PloS One*, 6(2), e17038. doi:10.1371/journal.pone.0017038

Kuska, B. (1998). Beer, Bethesda, and biology: how 'genomics' came into being. *JNCI Journal of the National Cancer Institute*, 90(2), 93–3. doi:10.1093/jnci/90.2.93

Kyrpides, N. C. (1999). Genomes OnLine Database (GOLD 1.0): a monitor of complete and ongoing genome projects world-wide. *Bioinformatics*, 15(9), 773–4. doi:10.1093/bioinformatics/15.9.773

Kyrpides, N. C., Hugenholtz, P., Eisen, J. A., Woyke, T., Göker, M., Parker, C. T., ... Klenk, H.-P. (2014). Genomic encyclopedia of Bacteria and Archaea: sequencing a myriad of type strains. *PLoS Biology*, 12(8), e1001920. doi:10.1371/journal.pbio.1001920

Lander, E. S., Linton, L. M., Birren, B., Nusbaum, C., Zody, M. C., Baldwin, J., ... Szustakowki, J. (2001). Initial sequencing and analysis of the human genome. *Nature*, 409(6822), 860–921. doi:10.1038/35057062

Landry, J. J. M., Pyl, P. T., Rausch, T., Zichner, T., Tekkedil, M. M., Stütz, A. M., ... Steinmetz, L. M. (2013). The genomic and transcriptomic landscape of a HeLa cell line. *G3* (Bethesda, Md.), 3(8), 1213–24. doi:10.1534/g3.113.005777

Lartigue, C., Glass, J. I., Alperovich, N., Pieper, R., Parmar, P. P., Hutchison, C. A, ... Venter, J. C. (2007). Genome transplantation in bacteria: changing one species to another. *Science* (New York, NY), 317(5838), 632–8. doi:10.1126/science.1144622

Leach, J. D. (2012). *Honor Thy Symbionts*. CreateSpace Independent Publishing Platform.

Leitch, I. J. (2007). Genome sizes through the ages. *Heredity*, 99(2), 121–2. doi:10.1038/sj.hdy.6800981

Lemon, K. P., Armitage, G. C., Relman, D. A., & Fischbach, M. A. (2012). Microbiota-targeted therapies: an ecological perspective. *Science Translational Medicine*, 4(137), 137rv5. doi:10.1126/scitranslmed.3004183

Levy, S., Sutton, G., Ng, P. C., Feuk, L., Halpern, A. L., Walenz, B. P., . . . Venter, J. C. (2007). The diploid genome sequence of an individual human. *PLoS Biology*, 5(10), e254. doi:10.1371/journal.pbio.0050254

Ley, R. E., Turnbaugh, P. J., Klein, S., & Gordon, J. I. (2006). Microbial ecology: human gut microbes associated with obesity. *Nature*, 444(7122), 1022–3. doi:10.1038/4441022a

Li, R., Fan, W., Tian, G., Zhu, H., He, L., Cai, J., . . . Yang, H. (2010). The sequence and *de novo* assembly of the giant panda genome. *Nature*, 463(7279), 311–17. doi:10.1038/nature08696

Liittschwager, D. (2012). *A World in One Cubic Foot: Portraits of Biodiversity*. Chicago: University Of Chicago Press.

Lupski, J. R. (2013). Genetics. Genome mosaicism—one human, multiple genomes. *Science* (New York, NY), 341(6144), 358–9. doi:10.1126/science.1239503

McCliment, E. A., Nelson, C. E., Carlson, C. A., Alldredge, A. L., Witting, J., & Amaral-Zettler, L. A. (2011). An all-taxon microbial inventory of the Moorea coral reef ecosystem. *ISME Journal*, 6(2), 309–19. doi:10.1038/ismej.2011.108

McFall-Ngai, M., Hadfield, M. G., Bosch, T. C. G., Carey, H. V, Domazet-Lošo, T., Douglas, A. E., . . . Wernegreen, J. J. (2013). Animals in a bacterial world, a new imperative for the life sciences. *Proceedings of the National Academy of Sciences of the United States of America*, 110(9), 3229–36. doi:10.1073/pnas.1218525110

McKusick, V., & Ruddle, F. (1987). A new discipline, a new name, a new journal. *Genomics*, 1, 1–2. Retrieved from <http://scholar.google.com/scholar?hl=en&btnG=Search&q=intitle:A+New+Discipline,+A+New+Name,+A+New+Journal#0>.

Mascarelli, A. (2014). Climate-change adaptation: designer reefs. *Nature*, 508(7497), 444–6. doi:10.1038/508444a

Mattick, J. S. (2007). A new paradigm for developmental biology. *Journal of Experimental Biology*, 210(Pt 9), 1526–47. doi:10.1242/jeb.005017

Mattick, J. S. (2011). The genomic foundation is shifting. *Science*, 331(6019), 874. Retrieved from <http://www.sciencemag.org/content/331/6019/874.1.short>.

Mattick, J. S., Taft, R. J., & Faulkner, G. J. (2010). A global view of genomic information—moving beyond the gene and the master regulator. *Trends in Genetics: TIG*, 26(1), 21–8. doi:10.1016/j.tig.2009.11.002

May, R. M. (1988). How many species are there on Earth? *Science* (New York, NY), 241(4872), 1441–9. doi:10.1126/science.241.4872.1441

May, R. M. (2010). Ecology. Tropical arthropod species, more or less? *Science* (New York, NY), 329(5987), 41–2. doi:10.1126/science.1191058

Meadow, J. F., Bateman, A. C., Herkert, K. M., O'Connor, T. K., & Green, J. L. (2013). Significant changes in the skin microbiome mediated by the sport of roller derby. *PeerJ*, 1: e53. doi:10.7717/peerj.53

Mendez, F. L., Krahn, T., Schrack, B., Krahn, A.-M., Veeramah, K. R., Woerner, A. E., . . . Hammer, M. F. (2013). An African American paternal lineage adds an extremely ancient root to the human Y chromosome phylogenetic tree. *American Journal of Human Genetics*, 92(3), 454–9. doi:10.1016/j.ajhg.2013.02.002

Meyer, M., Kircher, M., Gansauge, M.-T., Li, H., Racimo, F., Mallick, S., . . . Pääbo, S. (2012). A high-coverage genome sequence from an archaic Denisovan individual. *Science* (New York, NY), 338(6104), 222–6. doi:10.1126/science.1224344

Miller, W., Hayes, V. M., Ratan, A., Petersen, D. C., Wittekindt, N. E., Miller, J., . . . Schuster, S. C. (2011). Genetic diversity and population structure of the endangered marsupial Sarcophilus harrisii (Tasmanian devil). *Proceedings of the National Academy of Sciences of the United States of America*, 108(30), 12348–53. doi:10.1073/pnas.1102838108

Mole, B. (2013). FDA gets to grips with faeces. *Nature*, 498(7453), 147–8. doi:10.1038/498147a

Mora, C., Rollo, A., & Tittensor, D. P. (2013). Comment on 'Can we name Earth's species before they go extinct?'. *Science* (New York, NY), 341(6143), 237. doi:10.1126/science.1237254

Mora, C., Tittensor, D. P., Adl, S., Simpson, A. G. B., & Worm, B. (2011). How many species are there on Earth and in the ocean? *PLoS Biology*, 9(8), e1001127. doi:10.1371/journal.pbio.1001127

Nasser, W., Beres, S. B., Olsen, R. J., Dean, M. A., Rice, K. A., Long, S. W., . . . Musser, J. M. (2014). Evolutionary pathway to increased virulence and epidemic group A Streptococcus disease derived from 3,615 genome sequences. *Proceedings of the National Academy of Sciences of the United States of America*, 111(17), E1768–76. doi:10.1073/pnas.1403138111

Nelson, K. E., Weinstock, G. M., Highlander, S. K., Worley, K. C., Creasy, H. H., Wortman, J. R., . . . Zhu, D. (2010). A catalog of reference genomes from the human microbiome. *Science* (New York, NY), 328(5981), 994–9. doi:10.1126/science.1183605

Nissimov, J. I., Worthy, C. A, Rooks, P., Napier, J. A., Kimmance, S. A., Henn, M. R., . . . Allen, M. J. (2011). Draft genome sequence of the Coccolithovirus Emiliania huxleyi virus 203. *Journal of Virology*, 85(24), 13468–9. doi:10.1128/JVI.06440-11

Noble, D. (2008). Genes and causation. *Philosophical Transactions. Series A, Mathematical, Physical, and Engineering Sciences*, 366(1878), 3001–15. doi:10.1098/rsta.2008.0086

Nystedt, B., Street, N. R., Wetterbom, A., Zuccolo, A., Lin, Y.-C., Scofield, D. G., ... Jansson, S. (2013). The Norway spruce genome sequence and conifer genome evolution. *Nature*, 497(7451), 579–84. doi:10.1038/nature12211

O'Huallachain, M., Karczewski, K. J., Weissman, S. M., Urban, A. E., & Snyder, M. P. (2012). Extensive genetic variation in somatic human tissues. *Proceedings of the National Academy of Sciences of the United States of America*, 109(44), 18018–23. doi:10.1073/pnas.1213736109

O'Malley, M. A. (2008). 'Everything is everywhere: but the environment selects': ubiquitous distribution and ecological determinism in microbial biogeography. *Studies in History and Philosophy of Biological and Biomedical Sciences*, 39(3), 314–25. doi:10.1016/j.shpsc.2008.06.005

Oakeson, K. F., Gil, R., Clayton, A. L., Dunn, D. M., von Niederhausern, A. C., Hamil, C., ... Dale, C. (2014). Genome degeneration and adaptation in a nascent stage of symbiosis. *Genome Biology and Evolution*, 6(1), 76–93. doi:10.1093/gbe/evt210

Orlando, L., Ginolhac, A., Zhang, G., Froese, D., Albrechtsen, A., Stiller, M., ... Willerslev, E. (2013). Recalibrating Equus evolution using the genome sequence of an early Middle Pleistocene horse. *Nature*, 499(7456), 74–8. doi:10.1038/nature12323

Ottesen, E. A, Young, C. R., Eppley, J. M., Ryan, J. P., Chavez, F. P., Scholin, C. A., & DeLong, E. F. (2013). Pattern and synchrony of gene expression among sympatric marine microbial populations. *Proceedings of the National Academy of Sciences of the United States of America*, 110(6), E488–97. doi:10.1073/pnas.1222099110

Pagani, I., Liolios, K., Jansson, J., Chen, I.-M. A., Smirnova, T., Nosrat, B., ... Kyrpides, N. C. (2012). The Genomes OnLine Database (GOLD) v.4: status of genomic and metagenomic projects and their associated metadata. *Nucleic Acids Research*, 40(Database issue), D571–9. doi:10.1093/nar/gkr1100

Peng, Z., Lu, Y., Li, L., Zhao, Q., Feng, Q., Gao, Z., ... Jiang, Z. (2013). The draft genome of the fast-growing non-timber forest species moso bamboo (Phyllostachys heterocycla). *Nature Genetics*, 45(4), 456–61, 461e1–2. doi:10.1038/ng.2569

Peterson, J., Garges, S., Giovanni, M., McInnes, P., Wang, L., Schloss, J. A., ... Guyer, M. (2009). The NIH Human Microbiome Project. *Genome Research*, 19(12), 2317–23. doi:10.1101/gr.096651.109

Philippe, N., Legendre, M., Doutre, G., Couté, Y., Poirot, O., Lescot, M., ... Abergel, C. (2013). Pandoraviruses: amoeba viruses with genomes up to 2.5 Mb reaching that of parasitic eukaryotes. *Science* (New York, NY), 341(6143), 281–6. doi:10.1126/science.1239181

Pimm, S. L., Jenkins, C. N., Abell, R., Brooks, T. M., Gittleman, J. L., Joppa, L. N., ... Sexton, J. O. (2014). The biodiversity of species and their rates of extinction, distribution, and protection. *Science* (New York, NY), 344(6187), 1246752. doi:10.1126/science.1246752

Poinar, H. N., Schwarz, C., Qi, J., Shapiro, B., Macphee, R. D. E., Buigues, B., ... Schuster, S. C. (2006). Metagenomics to paleogenomics: large-scale sequencing of mammoth DNA. *Science* (New York, NY), 311(5759), 392–4. doi:10.1126/science.1123360

Purves, D., Scharlemann, J., Harfoot, M., Newbold, T., Tittensor, D. P., Hutton, J., & Emmott, S. (2013). Ecosystems: time to model all life on Earth. *Nature*, 493(7432), 295–7. doi:10.1038/493295a

Qin, J., Li, R., Raes, J., Arumugam, M., Burgdorf, K. S., Manichanh, C., ... Wang, J. (2010). A human gut microbial gene catalogue established by metagenomic sequencing. *Nature*, 464(7285), 59–65. doi:10.1038/nature08821

Rands, C. M., Darling, A., Fujita, M., Kong, L., Webster, M. T., Clabaut, C., ... Ponting, C. P. (2013). Insights into the evolution of Darwin's finches from comparative analysis of the Geospiza magnirostris genome sequence. *BMC Genomics*, 14(1), 95. doi:10.1186/1471-2164-14-95

Ratnasingham, S., & Hebert, P. D. N. (2007). BOLD: the Barcode of Life Data System (<http://www.barcodinglife.org>). *Molecular Ecology Notes*, 7(3), 355–64. doi:10.1111/j.1471-8286.2007.01678.x

Read, B. A, Kegel, J., Klute, M. J., Kuo, A., Lefebvre, S. C., Maumus, F., ... Grigoriev, I. V. (2013). Pan genome of the phytoplankton Emiliania underpins its global distribution. *Nature*, 499(7457), 209–13. doi:10.1038/nature12221

Reardon, S. (2014). Mugshots built from DNA data. *Nature News* (20 Mar.), 1–3. doi:10.1038/nature.2014.14899

Relman, D. A., & Falkow, S. (2001). The meaning and impact of the human genome sequence for microbiology. *Trends in Microbiology*, 9(5), 206–8. Retrieved from <http://www.ncbi.nlm.nih.gov/pubmed/11336835>.

Ridaura, V. K., Faith, J. J., Rey, F. E., Cheng, J., Duncan, A. E., Kau, A. L., ... Gordon, J. I. (2013). Gut microbiota from twins discordant for obesity modulate metabolism in mice. *Science* (New York, NY), 341(6150), 1241214. doi:10.1126/science.1241214

Rinke, C., Schwientek, P., Sczyrba, A., Ivanova, N. N., Anderson, I. J., Cheng, J.-F., ... Woyke, T. (2013). Insights into the phylogeny and coding potential of microbial dark matter. *Nature*, 499(7459), 431–7. doi:10.1038/nature12352

Rivière, J.-B., Mirzaa, G. M., O'Roak, B. J., Beddaoui, M., Alcantara, D., Conway, R. L., ... Dobyns, W. B. (2012). De novo germline and postzygotic mutations in AKT3, PIK3R2 and PIK3CA cause a spectrum of related megalencephaly syndromes. *Nature Genetics*, 44(8), 934–40. doi:10.1038/ng.2331

Robinson, G. E., Hackett, K. J., Purcell-Miramontes, M., Brown, S. J., Evans, J. D., Goldsmith, M. R., ... Schneider, D. J. (2011). Creating a buzz about insect genomes. *Science* (New York, NY), 331(6023), 1386. doi:10.1126/science.331.6023.1386

Roullier, C., Benoit, L., McKey, D. B., & Lebot, V. (2013). Historical collections reveal patterns of diffusion of sweet potato in Oceania obscured by modern plant movements and recombination. *Proceedings of the National Academy of*

Sciences of the United States of America, 110(6), 2205–10. doi:10.1073/pnas.1211049110

Rusch, D. B., Halpern, A. L., Sutton, G., Heidelberg, K. B., Williamson, S., Yooseph, S., . . . Venter, J. C. (2007). The *Sorcerer II* Global Ocean Sampling Expedition: northwest Atlantic through eastern tropical Pacific. *PLoS Biology*, 5(3), e77. doi:10.1371/journal.pbio.0050077

Sanger, F., Air, G. M., Barrell, B. G., Brown, N. L., Coulson, A. R., Fiddes, J. C., . . . Smith, M. (1977). Nucleotide sequence of bacteriophage ΦX174 DNA. *Nature*, 265(5596), 687–95. doi:10.1038/265687a0

Schnell, I. B., Thomsen, P. F., Wilkinson, N., Rasmussen, M., Jensen, L. R. D., Willerslev, E., . . . Gilbert, M. T. P. (2012). Screening mammal biodiversity using DNA from leeches. *Current Biology: CB*, 22(8), R262–3. doi:10.1016/j.cub.2012.02.058

Scholin, C. (2013). Ecogenomic sensors. In S. A. Levin (ed.), *Encyclopedia of Biodiversity* (vol. 2, pp. 690–700). Waltham, MA: Academic Press. doi:10.1016/B978-0-12-384719-5.00408-1

Scolari, F., Schetelig, M. F., Bertin, S., Malacrida, A. R., Gasperi, G., & Wimmer, E. A. (2008). Fluorescent sperm marking to improve the fight against the pest insect Ceratitis capitata (Wiedemann; Diptera: Tephritidae). *New Biotechnology*, 25(1), 76–84. doi:10.1016/j.nbt.2008.02.001

Shapiro, M. D., Kronenberg, Z., Li, C., Domyan, E. T., Pan, H., Campbell, M., . . . Wang, J. (2013). Genomic diversity and evolution of the head crest in the rock pigeon. *Science* (New York, NY), 339(6123), 1063–7. doi:10.1126/science.1230422

Shinzato, C., Shoguchi, E., Kawashima, T., Hamada, M., Hisata, K., Tanaka, M., . . . Satoh, N. (2011). Using the Acropora digitifera genome to understand coral responses to environmental change. *Nature*, 476(7360), 320–3. doi:10.1038/nature10249

Shoguchi, E., Shinzato, C., Kawashima, T., Gyoja, F., Mungpakdee, S., Koyanagi, R., . . . Satoh, N. (2013). Draft assembly of the Symbiodinium minutum nuclear genome reveals dinoflagellate gene structure. *Current Biology: CB*, 23(15), 1399–408. doi:10.1016/j.cub.2013.05.062

Shreeve, J. (2005). *The Genome War: How Craig Venter Tried to Capture the Code of Life and Save the World*. New York: Ballantine Books.

Shuster, S. M., Lonsdorf, E. V, Wimp, G. M., Bailey, J. K., & Whitham, T. G. (2006). Community heritability measures the evolutionary consequences of indirect genetic effects on community structure. *Evolution; International Journal of Organic Evolution*, 60(5), 991–1003. Retrieved from <http://www.ncbi.nlm.nih.gov/pubmed/16817539>.

Siddall, M. E., Kvist, S., Phillips, A., & Oceguera-Figuero, A. (2012). DNA barcoding of parasitic nematodes: is it kosher? *Journal of Parasitology*, 98(3), 692–4. doi:10.1645/GE-2994.1

Skloot, R. (2010). *The Immortal Life of Henrietta Lacks*. New York: Crown.

Smarr, L. (2012). Quantifying your body: a how-to guide from a systems biology perspective. *Biotechnology Journal*, 7(8), 980–91. doi:10.1002/biot.201100495

Smith, H. O., Hutchison, C. A., Pfannkoch, C., & Venter, J. C. (2003). Generating a synthetic genome by whole genome assembly: phiX174 bacteriophage from synthetic oligonucleotides. *Proceedings of the National Academy of Sciences of the United States of America*, 100(26), 15440–5. doi:10.1073/pnas.2237126100

Smith, M. B., Kelly, C., & Alm, E. J. (2014). Policy: How to regulate faecal transplants. *Nature*, 506(7488), 290–1. doi:10.1038/506290a

Sonnenburg, J. L. (2010). Microbiology: genetic pot luck. *Nature*, 464(7290), 837–8. doi:10.1038/464837a

Star, B., Nederbragt, A. J., Jentoft, S., Grimholt, U., Malmstrøm, M., Gregers, T. F., ... Jakobsen, K. S. (2011). The genome sequence of Atlantic cod reveals a unique immune system. *Nature*, 477(7363), 207–10. doi:10.1038/nature10342

Sthultz, C. M., Gehring, C. A., & Whitham, T. G. (2009). Deadly combination of genes and drought: increased mortality of herbivore-resistant trees in a foundation species. *Global Change Biology*, 15(8), 1949–61. doi:10.1111/j.1365-2486.2009.01901.x

Storey, A. A, Athens, J. S., Bryant, D., Carson, M., Emery, K., deFrance, S., ... Matisoo-Smith, E. (2012). Investigating the global dispersal of chickens in prehistory using ancient mitochondrial DNA signatures. *PloS One*, 7(7), e39171. doi:10.1371/journal.pone.0039171

Suzuki, T. A., & Worobey, M. (2014). Geographical variation of human gut microbial composition. *Biology Letters*, 10(2), 20131037. doi:10.1098/rsbl.2013.1037

Taft, R. J., Pheasant, M., & Mattick, J. S. (2007). The relationship between non-protein-coding DNA and eukaryotic complexity. *BioEssays: News and Reviews in Molecular, Cellular and Developmental Biology*, 29(3), 288–99. doi:10.1002/bies.20544

Tettelin, H., Masignani, V., Cieslewicz, M. J., Donati, C., Medini, D., Ward, N. L., ... Fraser, C. M. (2005). Genome analysis of multiple pathogenic isolates of *Streptococcus agalactiae*: implications for the microbial 'pan-genome'. *Proceedings of the National Academy of Sciences of the United States of America*, 102(39), 13950–5. doi:10.1073/pnas.0506758102

Toll-Riera, M., Bosch, N., Bellora, N., Castelo, R., Armengol, L., Estivill, X., & Albà, M. M. (2009). Origin of primate orphan genes: a comparative genomics approach. *Molecular Biology and Evolution*, 26(3), 603–12. doi:10.1093/molbev/msn281

Topol, E. (2013). *Creative Destruction of Medicine: How the Digital Revolution Will Create Better Health Care*. New York: Basic Books.

Topol, E. J. (2014). Individualized medicine from prewomb to tomb. *Cell*, 157(1), 241–53. doi:10.1016/j.cell.2014.02.012

Turnbaugh, P. J., Ley, R. E., Mahowald, M. A, Magrini, V., Mardis, E. R., & Gordon, J. I. (2006). An obesity-associated gut microbiome with increased capacity for energy harvest. *Nature*, 444(7122), 1027–31. doi:10.1038/nature05414

Tuskan, G. A, Difazio, S., Jansson, S., Bohlmann, J., Grigoriev, I., Hellsten, U., . . . Rokhsar, D. (2006). The genome of black cottonwood, Populus trichocarpa (Torr. & Gray). *Science* (New York, NY), 313(5793), 1596–604. doi:10.1126/science.1128691

Tyson, G. W., Chapman, J., Hugenholtz, P., Allen, E. E., Ram, R. J., Richardson, P. M., . . . Banfield, J. F. (2004). Community structure and metabolism through reconstruction of microbial genomes from the environment. *Nature*, 428(6978), 37–43. doi:10.1038/nature02340

Ursell, L. K., Van Treuren, W., Metcalf, J. L., Pirrung, M., Gewirtz, A., & Knight, R. (2013). Replenishing our defensive microbes. *BioEssays: News and Reviews in Molecular, Cellular and Developmental Biology*, 35(9), 810–17. doi:10.1002/bies.201300018

Van Valen, L. M., & Maiorana, V. C. (1991). HeLa, a new microbial species. *Evolutionary Theory & Review*, 10, 71–4.

Venter, J. C. (2013). *Life at the Speed of Light: From the Double Helix to the Dawn of Digital Life*. New York: Viking Adult.

Venter, J. C., Adams, M. D., Myers, E. W., Li, P. W., Mural, R. J., Sutton, G. G., . . . Zhu, X. (2001). The sequence of the human genome. *Science* (New York, NY), 291(5507), 1304–51. doi:10.1126/science.1058040

Venter, J. C., Remington, K., Heidelberg, J. F., Halpern, A. L., Rusch, D., Eisen, J. A., . . . Smith, H. O. (2004). Environmental genome shotgun sequencing of the Sargasso Sea. *Science* (New York, NY), 304(5667), 66–74. doi:10.1126/science.1093857

Venters, B. J., & Pugh, B. F. (2013). Genomic organization of human transcription initiation complexes. *Nature*, 502(7469), 53–8. doi:10.1038/nature12535

Watson, J., & Crick, F. (1953). A structure for deoxyribose nucleic acid. *Nature* 171(4356), 737–8. Retrieved from <http://symposium.cshlp.org/content/18/123.short>

Wheeler, D. A, Srinivasan, M., Egholm, M., Shen, Y., Chen, L., McGuire, A., . . . Rothberg, J. M. (2008). The complete genome of an individual by massively parallel DNA sequencing. *Nature*, 452(7189), 872–6. doi:10.1038/nature06884

Whitham, T. G., Difazio, S. P., Schweitzer, J. A, Shuster, S. M., Allan, G. J., Bailey, J. K., & Woolbright, S. A. (2008). Extending genomics to natural communities and ecosystems. *Science* (New York, NY), 320(5875), 492–5. doi:10.1126/science.1153918

Willerslev, E., Hansen, A. J., Binladen, J., Brand, T. B., Gilbert, M. T. P., Shapiro, B., . . . Cooper, A. (2003). Diverse plant and animal genetic records from Holocene and Pleistocene sediments. *Science* (New York, NY), 300(5620), 791–5. doi:10.1126/science.1084114

Woese, C. R., & Fox, G. E. (1977). Phylogenetic structure of the prokaryotic domain: the primary kingdoms. *Proceedings of the National Academy of Sciences of the United States of America,* 74(11), 5088–90. Retrieved from <http://www.pubmedcentral.nih.gov/articlerender.fcgi?artid=432104&tool=pmcentrez&rendertype=abstract>.

Woolhouse, M., & Farrar, J. (2014). Policy: an intergovernmental panel on antimicrobial resistance. *Nature,* 509(7502), 555–7. Retrieved from <http://www.ncbi.nlm.nih.gov/pubmed/24877180>.

Wu, D., Hugenholtz, P., Mavromatis, K., Pukall, R., Dalin, E., Ivanova, N. N., ... Eisen, J. A. (2009). A phylogeny-driven genomic encyclopaedia of Bacteria and Archaea. *Nature,* 462(7276), 1056–60. doi:10.1038/nature08656

Wulf, A. (2013). *Chasing Venus: The Race to Measure the Heavens.* Vintage.

Yong, E. (2013). Giant viruses open Pandora's box. *Nature News* (18 July), 1–4. doi:10.1038/nature.2013.13410

Zhang, G., Parker, P., Li, B., Li, H., & Wang, J. (2012). The genome of Darwin's finch (Geospiza fortis). *GigaScience,* 1(13). Retrieved from <http://dx.doi.org/10.5524/100040>.

Zhao, S., Zheng, P., Dong, S., Zhan, X., Wu, Q., Guo, X., ... Wei, F. (2013). Whole-genome sequencing of giant pandas provides insights into demographic history and local adaptation. *Nature Genetics,* 45(1), 67–71. doi:10.1038/ng.2494

Zhu, L., Wu, Q., Dai, J., Zhang, S., & Wei, F. (2011). Evidence of cellulose metabolism by the giant panda gut microbiome. *Proceedings of the National Academy of Sciences of the United States of America,* 108(43), 17714–9. doi:10.1073/pnas.1017956108

Zimmer, C. (2012). Tending the body's microbial garden. *New York Times* (18 June).

Zuccato, E., Castiglioni, S., Bagnati, R., Chiabrando, C., Grassi, P., & Fanelli, R. (2008). Illicit drugs, a novel group of environmental contaminants. *Water Research,* 42(4–5), 961–8. doi:10.1016/j.watres.2007.09.010

FURTHER READING

In addition to the endnotes and original references to the primary scientific literature found throughout the book, we recommend the 12 books listed at the end of this section as further reading. All are enjoyable, accessible, and widely read.

Like most biologists, our original scientific source of inspiration for this work is the *Origin of Species* (Darwin, 1859). We have also been inspired by the books of Richard Dawkins: *The Selfish Gene* (Dawkins, 1976) that placed DNA centre stage in the study of evolution, and his subsequent book on long-range genetic interactions, *The Extended Phenotype* (Dawkins, 1982) that presaged community genomics.

We believe that this is the first book to cover 'biodiversity genomics' as an emerging field, but many books have dealt with the topics of 'biodiversity' and 'genomics' separately. Among classic books on biodiversity, E. O. Wilson's *The Diversity of Life* (New York: W. W. Norton and Company, 1992) is perhaps the most relevant to our theme. Genomics is already a vast field and there are some excellent books available on the subject, particularly in relation to the Human Genome and its pioneers. The personal story of how Watson and Crick discovered the structure of DNA is recounted by Watson in his classic *The Double Helix* (Watson, 1968), while the race to sequence the human genome is told by Jamie Shreeve in *The Genome War* (Shreeve, 2005). Matt Ridley's *Genome: The Autobiography of a Species in 23 Chapters* (Ridley, 1999) is well worth reading. Ridley describes his book as a 'whistle-stop tour of some of the more interesting sites along the [human] genome and what they tell us about ourselves'.

Books on the emerging subfields of biocoding include those on personal genomics (Topol, 2013), synthetic biology (Church & Regis, 2012; Venter, 2013), microbiomics (Blaser, 2014; Knight & Buhler, 2015), and self-study (Duncan, 2009).

Blaser, M. J. (2014). *Missing Microbes: How the Overuse of Antibiotics Is Fueling Our Modern Plagues*. New York: Henry Holt and Co.

Church, G., & Regis, E. (2012). *Regenesis: How Synthetic Biology Will Reinvent Nature and Ourselves*. New York: Basic Books.

Darwin, C. R. (1859). *On the Origin of Species by Means of Natural Selection, or the Preservation of Favoured Races in the Struggle for Life*. London: John Murray. Retrieved from http://darwin-online.org.uk/converted/pdf/1859_Origin_F373.pdf

Dawkins, R. (1976). *The Selfish Gene*. New York: Oxford University Press.

Dawkins, R. (1982). *The Extended Phenotype: The Long Reach of the Gene*. Oxford: Oxford University Press.

Duncan, D. E. (2009). *Experimental Man: What One Man's Body Reveals about His Future, Your Health, and Our Toxic World*. Hoboken, NJ: Wiley.

Knight, R., & Buhler, B. (2015). *Follow Your Gut: The Enormous Impact of Tiny Microbes*. New York: Simon & Schuster/TED.

Ridley, M. (1999). *Genome: The Autobiography of a Species in 23 Chapters*. London: Fourth Estate.

Shreeve, J. (2005). *The Genome War: How Craig Venter Tried to Capture the Code of Life and Save the World*. New York: Ballantine Books.

Topol, E. (2013). *Creative Destruction of Medicine: How the Digital Revolution Will Create Better Health Care*. New York: Basic Books.

Venter, J. C. (2013). *Life at the Speed of Light: From the Double Helix to the Dawn of Digital Life*. New York: Viking Adult.

Watson, J. D. (1968). *The Double Helix: A Personal Account of the Discovery of the Structure of DNA*. New York: Scribner.

INDEX